FEAR OF FOOD

FEAR OF Food

A HISTORY OF

Why We Worry about What We Eat

Harvey Levenstein

The University of Chicago Press

CHICAGO AND LONDON

The University of Chicago Press, Chicago 60637
The University of Chicago Press, Ltd., London

Paperback edition 2013
Printed in the United States of America

22 21 20 19 18 17 16 15 14 13   5 6 7 8

ISBN-13: 978-0-226-47374-1 (cloth)
ISBN-13: 978-0-226-05490-2 (paper)
ISBN-13: 978-0-226-47373-4 (e-book)
DOI: 10.7208/chicago/9780226473734.001.0001

Library of Congress Cataloging-in-Publication Data

Levenstein, Harvey A., 1938–
Fear of food: a history of why we worry about what we eat / Harvey Levenstein.
p. cm.
Includes bibliographical references and index.
ISBN-13: 978-0-226-47374-1 (cloth: alkaline paper)
ISBN-10: 0-226-47374-0 (cloth: alkaline paper) 1. Food—United States—Psychological aspects. 2. Nutrition—United States—Psychological aspects. 3. Diet—United States. 4. Food preferences—United States. 5. Eating disorders—United States. 6. Phobias—United States. I. Title.
TX360.U6L47 2012
613.2—dc23

2011035700

⊗ This paper meets the requirements of ANSI/NISO Z39.48-1992 (Permanence of Paper).

Contents

Preface

The willingness to eat not for pleasure but for health is doubtless due to a fundamental U.S. trait: the fear of being sickly. Perhaps in England, but certainly not in France or Spain or Germany or Russia, will you find people so anxious to believe that by eating a certain way they can achieve a life buoyant and vigorous. Here it is the gourmet who is the curiosity, the dietitian who is the prophet.

Fortune, May 1936[1]

While writing two books on the history of American tourism to France, I became intrigued that so many of the Americans who toured France in the twentieth century had a very problematic relationship with French food. Although much of the world regarded the quality of French food as unparalleled, middle-class Americans often approached it with fear and trepidation. They worried that the famed French sauces camouflaged tainted meat, that the unfamiliar foods on the menus could make them sick, and that restaurant kitchens and servers were unhygienic. (This is not to mention the toilets!) Why, I began to wonder, did they have such a fraught relationship with food?[2]

Meanwhile, the American psychologist Paul Rozin and the French sociologist Claude Fischler were doing cross-national surveys of attitudes toward food that highlighted how different Americans were from Europeans in this regard. One survey showed that the French and Americans were at opposite ends of the spectrum on the scale of food fears. For instance, when asked what came to mind at the mention of whipped cream and chocolate, the French tended to respond with thoughts of pleasure, while Americans replied with words such as "guilt" or "unhealthy." This prompted Rozin to observe, "There is a sense among many Americans that food is as much a poison as it is a nutrient, and that eating is almost as dangerous as not eating." A later survey, co-directed by Fischler, reinforced this. It showed Americans to be more apprehensive and feel guiltier about food than people in France and four other European nations. The contrast was especially marked with the French, who tended to regard eating as a social, rather than an individual, act—that is, as a convivial pleasure. Americans, on the other hand, were plagued by a sense of individual responsibility in making their food choices, which were dictated by "dietetic" rather than "culinary" concerns.[3]

All of this spurred me to reexamine why Americans, who for much of the twentieth century regarded themselves as "the best-fed people in the world," had become this way. I say "reexamine," because in 1993 I concluded *Paradox of Plenty*, the second of my two histories of modern American food, with a reference to "the paradox of a people surrounded by abundance who are unable to enjoy it."[4] However, the main thrust of that book and its predecessor, *Revolution at the Table*, was to examine the wide range of forces that have shaped American food habits since about 1880. Fear of food played but a minor role in those works. In this book, I have used some of the topics mentioned in them, such as the various attempts to clean up the food supply and to combat the effects of large-scale processing, as jumping-off points for further research into how these contributed to Americans' anxieties about food. This has allowed me to look at topics such as the campaigns for "pure" and "natural" foods, Vitamania, and fear of saturated fat from a perspective that is different from other studies, including mine. The result, I hope, is a book that will appeal to readers interested in how Americans' current relationships to food developed over time. I hope as well that it might serve as an antidote to much of the current fearmongering about food. If it can help lessen even a few people's anxieties and increase the pleasure they get from eating, I will regard it as a success.

I must once again thank my wife, Mona, for being what the old radio comedian used to call "my boon companion and severest critic." As with my previous books, she went over the drafts of the manuscript with a sharp yet sympathetic eye, making many useful suggestions for revision. My friend Claude Fischler of the Centre national de la recherche scientifique (CNRS) in Paris read much of the manuscript from the perspective of a social scientist, while my molecular biologist friend Jack Pasternak, professor emeritus at the University of Waterloo, provided the perspective of a *real* scientist. Dr. Donald Rosenthal, professor emeritus at McMaster University's Michael G. DeGroote School of Medicine, helped reinforce my skepticism about the confidence with which medical wisdom is often dispensed and vetted parts of the manuscript for medical errors. Professor Paul Rozin of the University of Pennsylvania read the completed manuscript for the publisher and made a number of useful suggestions that I have tried to incorporate. As usual, Doug Mitchell, my editor at the University of Chicago Press, was very supportive in helping the manuscript over the various hurdles. Of course, its message about enjoying food went down well with such a dedicated *bec fin*, who allows no fears to impede his appreciation of the pleasures of the table. His *aide-de-camp*, Tim McGovern, was, as always, super-competent at helping to get the manuscript into publishable shape.

Parts of this book are based on the lectures I gave in 2009 at the annual Joanne Goodman Lecture Series of the University of Western Ontario in London, Ontario. The Joanne Goodman Lecture Series was established by Joanne's family and friends to perpetuate the memory of her blithe spirit, her quest for knowledge, and the rewarding years she spent at the University of Western Ontario.

Hamilton, Ontario, Canada
January 2011

Introduction

At the root of our anxiety about food lies something that is common to all humans—what Paul Rozin has called the "omnivore's dilemma."[1] This means that unlike, say, koala bears, whose diet consists only of eucalyptus leaves and who can therefore venture no further than where eucalyptus trees grow, our ability to eat a large variety of foods has enabled us to survive practically anywhere on the globe. The dilemma is that some of these foods can kill us, resulting in a natural anxiety about food. These days, our fears rest not on wariness about that new plant we just came across in the wild, but on fears about what has been done to our food before it reaches our tables. These are the natural result of the growth of a market economy that inserted middlemen between producers and consumers of food.[2] In recent years the ways in which industrialization and globalization have completely transformed how the food we eat is grown, shipped, processed, and sold have helped ratchet up these fears much further.

As a glance at a web page exposing "urban legends" will indicate, there are an amazing number of bizarre fears about our food supply (often involving Coca-Cola) always floating about. These might provide some insight into the nature of conspiracy theories, but they are not the kind of fear this book is about. What interests me are fears that have had the backing of the nation's most eminent scientific, medical, and governmental authorities. One of the protagonists in this book won the Nobel Prize; another thought he should have won it. Many of the others were the most eminent nutritional scientists of their time. The government agencies involved were staffed by experts at the top of their fields. Yet, as we shall see, many of the fears they stoked turned out to be either groundless or at best unduly exaggerated. Others involved frightening all Americans about things that should concern only a minority.

These scares, and the anxiety about food they have created, result from a confluence of forces that since the end of the nineteenth century have transformed how Americans eat and think about their food. First and foremost is something that contemporary home economists often noted: that for many years the production and preparation of food has been steadily migrating out of the home. In the country's early years, when 90 percent of Americans lived on farms, the outsiders handling their food were mainly millers and vendors of essentials like salt and molasses. One usually had a personal, trusting relationship with these suppliers, who were often neighbors. By the late nineteenth century, though, industrialization, urbanization, and a transportation revolution had transformed the nation. Cities boomed, railroads crisscrossed the nation, massive steamships crowded its ports, and urbanites were provided with foods that were not only not grown by neighbors—they were not even grown in neighboring countries. Large impersonal companies were now in charge of the canning, salting, refining, milling, baking, and other ways of preserving and preparing foods that had previously been done at home or by neighbors. All along the way, the foods passed through the hands of strangers with plenty of opportunities to profit by altering them, to the detriment of the quality and the healthfulness of the foods. There was, then, plenty of reason to mistrust what had happened to food before it reached the table.

These natural concerns were heightened by modern science. In the late nineteenth century, nutritional scientists discovered that food was not just undifferentiated fuel for the human engine. Rather, they said, it consisted of proteins, fats, and carbohydrates, each of which played a different role in preserving health. Only scientists could calculate how much of them were necessary. They then laid much of the groundwork for modern anxiety about food by warning that taste was the least reliable guide to healthy eating.

At the same time, fears were stoked by the new germ theory of disease, whose impact is discussed in the first two chapters of this book. The two chapters that follow deal with anxieties over the new chemicals that were used to preserve and process foods. There follows a chapter on how Americans' love affair with beef ended up giving it a kind of immunity from justifiable alarms. I then tell of how the discovery of vitamins gave rise to worries that modern food processing was removing these essential elements from foods. The following chapter tells the story of how concerns about processing led to the idealization of pre-industrial diets such as that of the inhabitants of a supposed Shangri-la in Pakistan and the consequent turn toward natural and organic foods. Finally, the last two chapters discuss lipophobia—the fear of dietary fat that some Americans have called "our national eating disorder."

Of course, it took much more than just some frightening ideas to arouse Americans about their food. The immense amounts of money involved in the food industries meant that, inevitably, huge financial stakes were involved as well. However, we shall see that the stakeholders were not just the usual suspects, the large corporations that dominated food production. They also included much less mendacious interests as well. Well-meaning public health authorities sought to demonstrate their importance by issuing exaggerated warnings about food dangers. Home economists helped justify their role in the education system by teaching how proper eating would avoid life-threatening diseases. During World War II, the federal government propagated the misguided notion that taking vitamins would make up for the deficiencies caused by food processing and help the nation defend itself from invasion. After the war, nonprofit philanthropies such as the American Heart Association raised billions of dollars in donations to spread the message that eating the wrong foods was killing millions of Americans. Scientific and medical researchers were awarded many more billions in government and corporate grants for studies warning about the dangers of eating fats, sugar, salt, and a host of other foods.

But the resulting food fears needed a receptive audience, and that is precisely what middle-class Americans were primed to be. By the early twentieth century, they had become the dominant force in American culture. Because they mainly lived in cities and large towns, they benefited most from the innovations in transportation, processing, and marketing that led to greatly expanded food choices. However, the resulting erosion of the reassuring personal relationships between sellers and buyers made them particularly susceptible to food scares. The media now became their major source of information about the safety of their food. Since much of this information was now scientific in origin, it was therefore the middle-class media—"quality" newspapers and magazines, and, later, radio and television news and public affairs shows—that played the major roles in disseminating it.

The residual Puritanism of the American middle class also helped make them susceptible to food fears. A culture that for hundreds of years encouraged people to feel guilty about self-indulgence, one that saw the road to salvation as paved by individual self-denial, made them particularly receptive to calls for self-sacrifice in the name of healthy living. This helped them lend a sympathetic ear to scientific nutritionists' repeated warnings that good taste—that is, pleasure—is the worst guide to healthy eating. By the end of the twentieth century, this guilt-ridden culture seemed to have weakened, as the notion that self-indulgence would benefit both individuals and society gained ground. However, at the heart of this more solipsistic view of life there still lay the old idea that illness and death were the result of an individual's own actions, including—and often especially—how they ate. ("Americans," a British wag has remarked, "like to think that death is optional.")

As a result, middle-class Americans have lurched from worrying about one fear of alimentary origin to another, with no apparent end in sight. As I write this, there is a burgeoning concern over salt in the diet. As with other such scares, experts are trying to frighten the entire nation about the health consequences of something that should concern only a minority (including, I might add, me). Typically, foods prepared and processed outside the home are taking most of the blame. Yet the demands of modern urban life seem to provide no easy way out of depending on such foods. Does this mean that anxiety about food will remain a recurring feature of middle-class life in North America? This study offers little hope that it will disappear. However, I would hope that by putting this anxiety into historical perspective, I might contribute in some small way to its diminution.

1
Germophobia

"Swat That Fly!"

That contemporary Americans fear food more than the French is rather ironic, for many modern food fears originated in France. It was there, in the 1870s, that the scientist Louis Pasteur transformed perceptions of illness by discovering that many serious diseases were caused by microscopic organisms called microbes, bacteria, or germs.[1] This "germ theory" of disease saved innumerable lives by leading to the development of a number of vaccines and the introduction of antiseptic procedures in hospitals. However, it also fueled growing fears about what industrialization was doing to the food supply.

Of course, fear of what unseen hands might be doing to our food is natural to omnivores, but taste, sight, smell (and the occasional catastrophic experience) were usually adequate for deciding what could or could not be eaten. The germ theory, however, helped remove these decisions from the realms of sensory perception and placed them in the hands of scientists in laboratories.

By the end of the nineteenth century, these scientists were using powerful new microscopes to paint an ever more frightening picture. First, they confirmed that germs were so tiny that there was absolutely no way that they could be detected outside a laboratory. In 1895 the *New York Times* reported that if a quarter of a million of one kind of these pathogenic bacteria were laid side by side, they would only take up an inch of space. Eight billion of another variety could be packed into a drop of fluid. Worse, their ability to reproduce was nothing short of astounding. There was a bacillus, it said, that in only five days could multiply quickly enough to fill all the space occupied by the waters of Earth's oceans.[2]

The reported dangers of ingesting germs multiplied exponentially as well. By 1900 Pasteur and his successors had shown that germs were the cause of deadly diseases such as rabies, diphtheria, and tuberculosis. They then became prime suspects in many ailments, such as cancer and smallpox, for which they were innocent. In 1902 a U.S. government scientist even claimed to have discovered that laziness was caused by germs.[3] Some years later Harvey Wiley, head of the government's Bureau of Chemistry, used the germ theory to explain why his bald head had suddenly produced a full growth of wavy hair. He had discovered, he said, that baldness was caused by germs in the scalp and had conquered it by riding around Washington, D.C., in his open car, exposing his head to the sun, which killed the germs.[4]

America's doctors were initially slow to adopt the germ theory, but public health authorities accepted it quite readily.[5] In the mid-nineteenth century, their movement to clean up the nation's cities was grounded in the theory that disease was spread by invisible miasmas—noxious fumes emanating from putrefying garbage and other rotting organic matter. It was but a short step from there to accepting the notion that dirt and garbage were ideal breeding grounds for invisible germs. Indeed, for a while the two theories coexisted quite happily, for it was initially thought that bacteria flourished only in decaying and putrefying substances—the very things that produced miasmas. It was not difficult, then, to accept the idea that foul-smelling toilets, drains, and the huge piles of horse manure that lined city streets harbored dangerous bacteria instead of miasmas. Soon germs became even more frightening than miasmas. Scientists warned that they were "practically ubiquitous" and were carried to humans in dust, in dirty clothing, and especially in food and beverages.[6]

The idea that dirt caused disease was accepted quite easily by middle-class Americans. They had been developing a penchant for personal cleanliness since early in the nineteenth century. Intoning popular notions such as "cleanliness is next to godliness," they had reinforced their sense of moral superiority over the "great unwashed" masses by bathing regularly and taking pride in the cleanliness of their houses. It was also embraced by the women teaching the new "domestic science" in the schools, who used it to buttress their shaky claims to be scientific. They could now teach "bacteriology in the kitchen," which meant learning "the difference between apparent cleanliness and chemical cleanliness."[7]

The nation's ubiquitous hucksters popularized the germ theory by selling potions said to kill germs in the bloodstream. Even divine intervention seemed to offer no protection against these invaders. A newspaper article promoting one such potion showed an assassin about to plunge a long dagger into the back of a man praying at a church altar. The headline warned, "Death Is Everywhere. Disease Germs Are Even More Deadly than the Assassin's Dagger. No One Is Safe. No Place Is Sacred."[8]

Food and beverages were thought to be particularly dangerous because they were the main vehicles for germs entering the body. As early as 1893, domestic scientists were warning that the fresh fruits and vegetables on grocers' shelves "quickly catch the dust of the streets, which we know is laden with germs possibly malignant."[9] A U.S. government food bulletin warned that dust was "a disease breeder" that infected food in the house and on the street.[10] In 1902 sanitation officials in New York City calculated that the air in the garbage-strewn streets of the crowded, impoverished Lower East Side harbored almost 2,000 times as many bacilli as the air on wealthy East Sixty-Eighth Street and warned that these airborne germs almost certainly infected the food in the hundreds of pushcarts that lined its streets.[11] The horse dung that was pulverized by vehicles on paved city streets was particularly irksome, blowing into people's faces and homes and covering the food merchants' outdoor displays. In 1908 a sanitation expert estimated that each year 20,000 New Yorkers died from "maladies that fly in the dust, created mainly by horse manure."[12]

The notion that dirty hands could spread germs, so important in saving lives in hospital operating rooms, was of course applied to food. In 1907 the nation was riveted by the story of "Typhoid Mary," a typhoid-carrying cook who infected fifty-three people who ate her food, three of whom died. She resolutely refused to admit that she was a carrier of the disease-causing germ and was ultimately institutionalized.[13] In 1912 Elie Metchnikoff, Pasteur's successor as head of the Pasteur Institute, announced that he had discovered that gastroenteritis, which killed an estimated 10,000 children a year in France and probably more in the United States, was caused by a microorganism found in fruits, vegetables, butter, and cheese that was transmitted to infants by mothers who had failed to wash their hands with soap before handling or feeding their babies. By then, any kind of dirt was thought to be synonymous with disease. "Wherever you find filth, you find disease," declared a prominent New York pathologist.[14]

New microscopes that made it easier to photograph bacteria revealed

DEATH IS EVERYWHERE.

DISEASE GERMS ARE EVEN MORE DEADLY THAN THE ASSASSIN'S DAGGER.

NO ONE IS SAFE. NO PLACE IS SACRED.

Disease claims the just and the unjust. It is no respecter of persons or places. It is just as likely to overtake the devout man at the altar as the ruffian in the disgraceful dive. The germs are everywhere—in the air we breathe—in the food we eat—in the water we drink. There is no way of escaping them except by keeping perfectly strong and well. A germ can find no lodging in a healthy body. Pure blood is an impenetrable armor against disease. A perfectly pure, healthy body has no place in which a germ may lodge. Such a body may come in contact with thousands of germs and never be harmed by them. Germs and surgeon of the Invalids' Hotel and Surgical Institute of Buffalo, N. Y., perhaps the most important institution of its kind in the world. For more than a quarter of a century this wonderful medicine has been doing its work of healing all over the world and thousands upon thousands of grateful people who have been made well by it have written the most glowing and grateful letters to its discoverer. It is a secret preparation, in the sense that its discoverer has not publicly divulged its ingredients. This has naturally made it unpopular with physicians whose patients it has cured after they had failed.

There has been a good reason for preserving this secrecy. If the ingredients were known, many people would believe that they could make a preparation just as good by using the same component parts. This is not true, for the care in preparation, and in procuring exactly the proper quantity of drugs, has perhaps as much to do with the efficiency of the medicine as the character of its ingredients. As it is, the wonderful success of the "Golden Medical Discovery" has caused it to be imitated, and has caused the preparation of many articles said to be similar. There is nothing that in any way approaches it in certainty and efficiency. The druggist who offers you something else, which he says is "just as good" is not an honest man. He is not a safe man to deal with. If he tries to give you something else in place of what you want, how do you know that he will fill your doctor's prescription properly? If he will en-

"Death Is Everywhere" reads this 1897 advertisement for a germ killer. It shows a man standing at a church altar and says: "Disease Germs Are Even More Deadly than the Assassin's Dagger. *No One Is Safe. No Place Is Sacred.*" (*Brooklyn Daily Eagle*, 1897)

even more foods infested with germs. A ditty called "Some Little Bug Is Going to Find You" became a hit. Its first verse was

> In these days of indigestion, it is oftentimes a question
> As to what to leave alone.
> Every microbe and bacillus has a different way to kill us
> And in time will claim us for their own
> There are germs of every kind in every food that you will find
> In the market and upon the bill of fare
> Drinking water is just as risky as the so-called "deadly" whiskey
> And it's often a mistake to breathe the air.[15]

Even house pets were caught in the dragnet. In 1912 when tests of cats' whiskers and fur in Chicago revealed the presence of alarming numbers of bacteria, the city's Board of Health issued a warning that cats were "extremely dangerous to humanity." A local doctor invented a cat trap to be planted in backyards to capture and poison wayward cats. The Topeka, Kansas, Board of Health issued an order that all cats must be sheared or killed.[16] The federal government refused to go that far but did warn housewives that pets carried germs to food, leaving it up to the families to decide what to do about them. What some of them did was to panic. A serious outbreak of infantile paralysis (polio) in New York City in 1916 led thousands of people to send their cats and dogs to be gassed by the Society for Prevention of Cruelty to Animals, despite the protestations of the city health commissioner that they did not carry the offending pathogen. From July 1 to July 26, over 80,000 pets were turned in for destruction, 10 percent of whom were dogs.[17]

But the most fearsome enemy in the war against germs was not a lovable pet, but the annoying housefly. Dr. Walter Reed's investigations of disease among the American troops who invaded Cuba during the Spanish-American War in 1898 had famously led to the discovery that mosquitoes carried and spread the germs that caused yellow fever. But yellow fever was hardly present in the United States. Much more important was his subsequent discovery that houseflies could carry the bacteria causing typhoid to food, for that disease killed an estimated 50,000 Americans a year.[18] Although Reed's studies had shown that this could happen only when flies were practically immersed in human, not animal, excrement, his observations soon metamorphosed into the belief that during the war typhoid-carrying flies had killed more American soldiers than had the Spanish.[19] This was buttressed by a kind of primitive epidemiology, wherein experts noted that typhoid peaked in the autumn, when the fly population was also at its peak.[20]

The number of diseases blamed on flies carrying germs to food then multiplied almost as rapidly as the creatures themselves. In 1905 the journal *American Medicine* warned that flies carried the tuberculosis bacillus and recommended that all food supplies be protected by screens.[21] That December a public health activist accused flies of spreading tuberculosis, typhoid, and a host of other diseases. "Suppose a fly merely walks over some fruits or cooked meats exposed for sale and [infects] them," he said. "If the food is not cooked to kill them the person who eats them is sure to have typhoid." The New York State Entomologist pointed out that during the warm months "the abundance of flies" coincided with a steep rise in deaths from typhoid and infant stomach ailments. "Nothing but criminal indifference or inexcusable ignorance is responsible for the swarms of flies so prevalent in many public places." A *New York Times* editorial agreed, saying, "It is cruelty to children to allow flies to carry death and disease wherever they go." Flies, the paper contended, "bear into the homes and food, water, and milk supplies of the people not only the germs of typhoid and cholera, but of tuberculosis, anthrax, diphtheria, opthalmia, smallpox, staphylococcus infection, swine fever, tropical sore, and the eggs of parasitic worms." A Philadelphia doctor opined that flies likely infected food with the germs that caused cancer.[22]

The December 1905 *Times* piece had ended by saying that it was no wonder that "sanitarians are now beginning a warfare against the fly."[23] And indeed, the first shots in that war had been fired earlier that year in Topeka, the capital city of Kansas, by Dr. Samuel Crumbine, the state director of public health. Crumbine was a short man who had begun his career by ostentatiously toting a large six-shooter through the violent streets of Dodge City, the legendary cow town. Now, in Topeka, he began putting up "Fly Posters" and distributing a "Fly Bulletin" warning of the dangers posed by the insect. His campaign bore little fruit until, at the opening game of the Topeka baseball team's 1905 season, inspiration struck. After hearing fans calling for hitters to "swat the ball" and hit a "sacrifice fly," he leaped to his feet, shouting, "I've got it!" He meant not a foul ball, but "Swat the Fly," which became the slogan that galvanized his campaign. A local schoolteacher then came to him with a brilliant new weapon, a yardstick with a square of wire mesh attached to its end that he called a "fly bat." Crumbine immediately rechristened it a "fly swatter," and the rest, as they say, is history.

Crumbine then enlisted Kansas schoolchildren as mercenaries in his war: they were paid in cash or movie tickets for bringing piles of dead flies to school. In ten weeks they presented their teachers with thirty bushels, or an estimated 7,172,000 flies. Crumbine's campaign, and the flyswatter, drew national attention, and for years thereafter many of the anti-fly campaigns that swept the nation adopted the slogan "Swat the Fly."[24]

The war on flies lasted for over ten years and was fought on many fronts—swatting flies, strewing poisons about, hanging flypaper in stores and houses, and using screens (another new invention) to cover windows, doors, and food itself. Using the slogan "Dirt Fattens Flies and Kills Babies," New York City authorities warned that "the little fly on the wall" spread the germs of typhoid, tuberculosis, and other diseases when it landed on food. They said it was "more dangerous than the tiger or cobra, and may easily be classed as the most dangerous animal on earth." The Massachusetts Board of Health warned against allowing the "filthy insects" near food, and the city of Boston appropriated $100,000—then a very large sum—for their eradication. Women in Rochester, New York, petitioned their government to do the same. Nurses' associations, university scientists, Boy Scouts, and businessmen joined the effort. In 1911 Edward Hatch, a member of the New York City merchants' anti-fly committee, commissioned a photographer to go to England to take pictures of "flies and their dangerous activities" through a new kind of microscope. Upon his return, the pictures were assembled into a movie that was shown all over the country. The *New York Times* praised it for bringing "millions of people to realize how great is the danger from flies." Socialites spending that summer in Kennebunkport, Maine, distributed hundreds of free tickets to the Bijou Theater, which showed nothing but "The Fly Film" for an entire week.[25]

By the summer of 1912, hundreds of communities were engaged in what was called "The Greatest Anti-Fly Crusade Ever Known." Federal government publications labeled it "the typhoid fly." Doctors now held it responsible for spreading the germs that caused tuberculosis, cholera infantum, gastroenteritis, spinal meningitis, and infantile paralysis. Home economists urged housewives to keep these "germs with wings" off their food. State boards of health, county commissions, municipal governments, private companies, and women's associations gave detailed instructions on how to poison, trap, swat, or otherwise bring flies' lives to a quick end. A pamphlet told Ohio schoolchildren that flies are "the most deadly enemy of man. They kill more people than all the lions, tigers, snakes, and even wars." The state of North Dakota called its fly-killing instructions "A Fly Catechism." Indiana and other states put up posters with colored cartoons for those who could not read.[26]

Many communities took a leaf from the Kansas book and offered monetary bounties to people who brought in the largest number of dead flies. One enterprising twelve-year-old impressed the editors of a Worcester, Massachusetts, newspaper by presenting them an astounding ninety-five quarts of flies, walking off with their $100 prize. Only later was it discovered that he had bred the flies himself in heaps of fish offal. Yet the zealous crusaders refused to let this fraud deter them. After all, said the *New York Times*, "Those at the head of the movement realized that children . . . would kill off more flies than the parents would have the time to find." So, schoolchildren throughout the country continued to be offered cash prizes for bringing in dead flies, but with a new proviso: time limits were set, amounting to less time than it took for flies to gestate and breed.[27]

The enterprising young boy in Worcester was by no means the only one who sought to profit from the campaign. Flypaper was advertised as a lifesaving device. There was a surge in sales of dubious disinfectants, such as Electrozone: sea water zapped by electrical currents that was to be rubbed into the skin and onto foods.[28] But those who benefited most from the wave of germophobia were the large food processors. They now promoted their mechanized production systems as preventing food from being touched by germ-laden human hands. Gold Medal flour advertised that "the hands of the miller have not come into contact with the food at any stage in its manufacture." Cardboard packages were now called "sanitary boxes."[29] Uneeda biscuits, wrapped in waxed paper and packed in cardboard boxes, rapidly put an end to the old cracker barrels, buzzing with flies, that had been a mainstay of American general stores. Their manufacturer, the National Biscuit Company, advertised that "Uneeda Biscuits are touched only once by human hands—when the pretty girls pack them."[30] Kellogg's cereals had forty-eight public health authorities attest that its "Waxtite" packaging prevented contamination by germs.[31] The H.J. Heinz Company called its products "the exemplification of purity," prepared in "model kitchens" by "neat uniformed workers." To prove it, the company opened its assembly lines to visitors and allowed them to watch women in white frocks stuff pickles into bottles. Even beer drinkers were reassured. The Schlitz Brewing Company said its beer was "brewed in absolute cleanliness and cooled in filtered air." It was then "sterilized after the bottle was sealed." Because of its purity, it said, "when your physician prescribes beer, it is always Schlitz beer."[32]

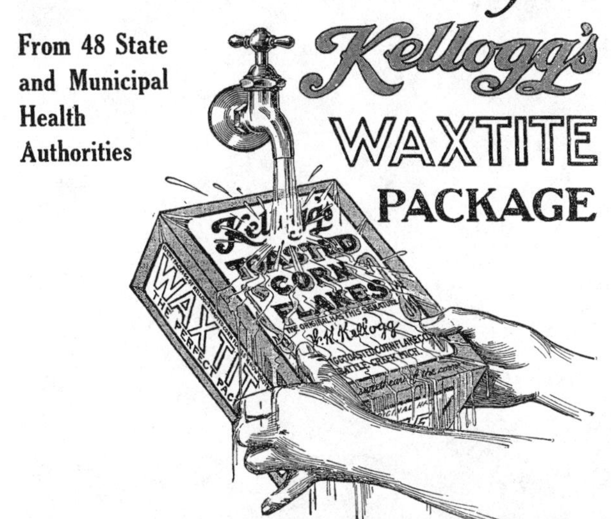

In 1914, with a so-called "War on Flies" raging, Kellogg's cited public health officials' assurances that its packaging provided protection from these dangerous germ-carriers.

In the early 1900s, canned foods were often regarded with suspicion, and with good reason. Many were produced by children in small, unsanitary facilities such as this. (Lewis Hine, Library of Congress)

The long-term impact of the "War on Flies" is reflected in the World War II government poster called "Never Give a Germ a Break!"

There was, of course, considerable irony in these claims, coming as they did from the same kind of sources that, by manipulating food far from consumers' eyes, were at the root of the underlying disquiet over the food supply. But the companies' success in allaying fear of germs proved to be an important step in making brand names invaluable assets in the food industries. From then on, the main prizes in food industry mergers and acquisitions were not companies' production facilities but their reassuring brand names.

The war against flies continued into 1917, even after that April, when the United States entered a war that was really killing people. Indeed, during the summer of that year, the *New York Times*, rallying the populace for the annual "Swat the Fly" movement, warned that in the Boer and Spanish-American wars combined, flies had "killed more men than bullets."[33] But by then the burgeoning fly population was being reduced not by schoolboys with flyswatters, but by the producer of a new kind of pollutant, the internal combustion engine. The growing number of automobiles on city streets helped dry out much of the horse manure that flies thrived on.[34] By 1917 cars had replaced horses to such an extent that the curbside piles of manure that provided such enticing dining-out opportunities for flies disappeared. They were soon being credited with practically eliminating the fly population on city streets.[35]

By then, though, there was also a growing realization that the bacteria the flies carried were not at all as dangerous as had been thought. "The truth is," said a 1920 summation of current medical opinion, "that you cannot stamp out bacteria. They are everywhere. . . . And, here is the point of the revolution—for the most part they are harmless."[36]

Yet the idea that flies carry dangerous germs remained permanently rooted in the national psyche. In the 1920s the U.S. Department of Agriculture expended considerable effort on teaching rural people diabolical ways to kill them en masse.[37] In the late 1940s, flies were wrongly blamed for causing a nationwide outbreak of polio, and whole towns and neighborhoods were sprayed with DDT to eradicate them. Even today most of us continue to associate flies with harmful germs. How many of us can eat food that we have seen covered in flies? And how many of us who have traveled in the "developing world" have not marveled at the insouciance of local vendors and shoppers, whose ancestors never experienced germophobia, to exactly that sight?

2

Milk: "The Most Valuable and Dangerous Food"

One of the most damning charges against flies was that they contaminated milk.[1] But this was just one of the fears that the germ theory had set to buzzing over cow's milk. Suspicions had often darkened views of the white liquid before then. Indeed, its reputation as a benign, healthful food was a relatively recent one. For much of the nineteenth century, a lot of the milk drunk by urban dwellers was unappetizing "swill milk," produced by sickly cows who were fed on the pungent effluent from distilleries. As a result, few adults drank fresh milk, especially in the summer, when it could sour within a few hours. However, in the 1880s and 1890s, refrigerated railroad cars began bringing cool fresh milk from dairy farms in the countryside, elevating the quality of milk sold in the cities enormously and helping to dissipate many traditional fears.

The fresh milk from the countryside was particularly welcomed by mothers. Employed working-class ones, who found it impossible to breast-feed their infants at the recommended two- or three-hour intervals, bottled it to feed their babies in their absence. Since cow's milk has too much fat for infants to digest, they would dilute it with water and add a bit of sugar to make it palatable to their children. Middle- and upper-class women who could not or chose not to breast-feed turned to the new milk supply as an alternative to the traditional wet nurses. Nutritional scientists devised powders for them to add to either fresh or condensed milk, along with water, to replicate the fat, protein, and carbohydrate content of mother's milk. Then pediatricians, whose clients were mainly the best-off mothers, came up with "percentage feeding." This involved them writing a prescription for a "formula" they said was tailored to each baby's particular needs. Mothers would then take these scripts calling for specific percentages of milk, water, and sugar to special milk laboratories to be filled.[2]

But just as fresh milk from farms was becoming more available and acceptable, its safety began to be questioned. First came the fear of typhoid. In the 1880s and 1890s, typhoid outbreaks had been commonly blamed on typhoid microbes in public water supplies.[3] At times typhoid was traced to milk, but the blame was placed not on the milk, but on greedy farmers and milk peddlers who diluted it with dirty water. In 1889 a scientist warned that the milk delivered to people's doors each morning carried "millions of living insects," called "microorganisms, or bacteria," that came from diluting the milk with water polluted by "drainage from the sick chamber."[4]

However, in the 1890s, even though most urban water supplies had been cleaned up, killer epidemics of infant diarrhea recurred with ever-greater ferocity, particularly in the summers. As a result, milk itself became the prime suspect.[5] By 1900 new microscopes were confirming this, showing that milk was a much friendlier host to typhoid germs than was water. Scientists now reevaluated the typhoid epidemics, concluding that many of them could be "traced, beyond a doubt, to the milk supply."[6] In 1903 the *New York Times* reported that a quart of milk "of average quality" contained over 12 million bacilli when fresh from the cow, but that within twenty-four hours these proliferated to close to 600 million.[7] A scientist now warned that "milk of all foods is most liable to be affected by impurities," particularly the tuberculosis and typhoid bacilli.[8]

Then came the discovery that the bacillus that caused tuberculosis, the deadly "white plague," infected not just cow's milk, but many of the cows themselves. The *New York Times* said that it was now clear that, in addition "to the deaths of thousands of babies" each year, cow's milk was responsible for "those of tens of thousands of others who die annually from 'the great white plague' after years of suffering."[9]

The indictment continued to expand. An investigation of a typhoid outbreak in Washington, D.C.'s milk supply concluded that not only was milk a natural carrier of typhoid; it also harbored microbes causing a number of other diseases. When the germs causing dreaded diphtheria and scarlet fever were added to the list, the *New York Times* said that milk was justifiably called "the most valuable and the most dangerous food."[10] The public health specialist Milton Rosenau did not waste words on its value. He warned that milk was "*apt to be dangerous to health.*"[11]

On the surface, the solution seemed simple: pasteurization. In the 1870s Louis Pasteur had discovered that slowly raising the temperature of a liquid to 165 degrees Fahrenheit, keeping it there for twenty minutes, and then rapidly cooling it would kill milk-borne microbes without affecting its palatability. In 1893 the philanthropist Nathan Straus set up a pasteurization facility to provide pasteurized milk for poor people in New York City's Lower East Side. By 1902 there were thirteen of them, dispensing close to 1 million bottles of free milk each summer.[12] But those who pushed for compulsory pasteurization found themselves thwarted by public health officials such as Harvey W. Wiley, chief of the U.S. Bureau of Chemistry, who argued (wrongly) that pasteurization depleted milk of its nutritional qualities. It was also opposed by those who said that pasteurization would make milk too expensive for the poor.[13] The thousands of milkmen who plied the city streets in their wagons, ladling fresh milk from large cans, joined the opposition. They said that compulsory pasteurization would put them out of business because they could not afford the expensive machines and bottles.[14]

Pasteurization's main foes, though, were advocates of a competing system, milk certification, which was even more expensive. It sought to verify that milk was safe by monitoring bacteria levels all the way from

In the early 1900s, unsanitary milk was blamed for the era's horrific levels of infant and child mortality. "Milk stations" were set up in cities to provide clean milk for the babies of the poor. Mrs. Adolph Strauss, whose financier husband funded many of them, poses here with some officials and the babies they helped. (Library of Congress)

the farmer's milk pail to the consumer's bottle. This was particularly favored by pediatricians, whose system of "percentage feeding" called for its use. However, although certification seemed to work well enough for the upper classes, who could afford pediatricians and expensive infant food, it was well out of reach of most people, and only 1 percent of the nation's milk supply was certified.[15] (Another alternative never got off the ground: Harvey Wiley proposed criminalizing improper infant feeding and imposing fines on mothers who gave their children impure milk.)[16]

The battle between the competing systems heightened fear of milk by exposing the public to horror stories regarding what both sides agreed on: that the urban milk supply was full of deadly germs and governments were doing little about it. One study found that 17 percent of the milk sampled in New York City's restaurants harbored tuberculosis germs. Yet in 1907 there were only fourteen inspectors in all of New York State who were responsible for weeding out an estimated 300,000 tubercular cows. Anti-fly crusaders in New York City said that the germs that caused 49,000 infants to die each year from gastroenteritis were "probably all carried to the milk by flies." However, New York City had only sixteen inspectors overseeing sanitary conditions in the 12,000 locales where milk was sold.[17]

The nation's doctors seemed about evenly divided over what to do. About half of them seemed to blame raw or certified milk for the astronomical rates of child and infant mortality, while the other half blamed pasteurized milk. Appeals to President Taft to appoint a commission to decide who was right failed, as the president, perhaps realizing that its conclusions would certainly enrage half of the nation's doctors, demurred.[18] Yet the accusations continued to mount. In 1909, 4,000 of the 16,000 deaths of infants under one year old in New York City were ascribed to "bad milk and feeding."[19]

It comes as no surprise, then, to learn that in the years from 1909 to 1916 American milk consumption dropped by almost 20 percent. Since from half to three-quarters of the milk produced was used in rural areas to make butter, cheese, and cream, that would mean that urban milk consumption plummeted by 40 percent to 60 percent—a rather amazing drop. Then, as new methods made large-scale pasteurization possible and advocates of laws making it mandatory began to win the day, consumption rose again. By 1920, when most of the urban milk supply was pasteurized, consumption was back at prewar levels.[20]

"The Perfect Food"

Although pasteurization was mainly responsible for dissipating the public's fear of milk, the way in which it transformed the milk industry also played an important role.[21] The small peddlers who had opposed it out of fear that they would be put out of business by their inability to buy expensive equipment and large numbers of bottles turned out to be right. In city after city, a few well-financed companies rapidly came to dominate milk distribution. Then, in the mid-1920s, the scale of consolidation rose again, as two giants, Borden's and National Dairy Products (Sealtest), embarked on massive takeover campaigns that gave them dominance over the national retail market. (In one month Borden's alone bought out thirty-nine companies.) At the same time, the actual production of milk came to be controlled by a small number of powerful producers' cooperatives in key dairy-producing states. These two groups, private and cooperative, gained government support in mounting a massive effort that helped complete milk's transformation from "the most dangerous food" into "the perfect food."

The process began in 1919, when the U.S. Department of Agriculture and the industry's National Dairy Council joined together to promote milk consumption in the schools with weeklong "milk for health" programs. These set the pattern for government and industry to flood the schools with posters, publications, plays, and songs telling of how milk prevented disease.[22] In 1921 the New York State Dairymen's League began providing a glass of milk free to every schoolchild, along with a letter for them to take home to their mothers in which public health authorities told them that "all doctors" said they should drink at least one and a half pints every day. School trips were arranged to nearby dairy farms, where they could see the sanitary conditions in which it was produced.[23] In New England the regional Dairy Council had teachers introduce students to "the Good Health Fairy," who told them that drinking four glasses of milk a day would provide all of the nutrients they needed for good health.[24]

The Dairy Council sponsored billboards and magazine advertisements featuring pictures of healthy-looking children happily sipping their milk. In the high schools, home economists extolled milk as essential for health, showing their students industry-supplied pictures of scrawny rats who had been raised on vegetable oil and sleek, healthy-looking ones who had been brought up on butter. Producers of cocoa and other sweet additives climbed aboard, advertising that doctors "prescribed" their products to help children drink their milk.[25]

The real breakthrough, though, resulted from promoting milk as essential for the health of adults. In 1921 the New York Dairymen's League persuaded the mayor of New York City to proclaim "Milk Week," during which he and the city's commissioner of health demonstrated their belief in its healthful properties by drinking a quart of it every day for lunch. Liggett's, a major drugstore chain, set up special "milk bars," at which busy businessmen could follow their example and quaff a quart of the healthy beverage as part of their quick lunches.[26]

Nutritionists lent an important hand to this remarkable turnabout. First, they upped their recommendation for how much children should drink from one and a half pints to one quart a day. Then they began calling on adults to drink the same amount. The vitamin researcher Elmer McCollum, the nation's most famous nutritional scientist, declared that the reason the Western diet was superior to the "Oriental" one was that "Orientals" drank little milk after weaning, as a result of which they were shorter and less vigorous. He warned, though, that since the beginning of the Industrial Revolution, consumption of milk and dairy products had been declining in the West. As a result, he said, there was "a pronounced tendency to physical deterioration." This stood in marked contrast to the healthfulness of "vigorous, aggressive . . . pastoral nomads" such as contemporary Arabs. It was because they consumed such "liberal amounts of milk" that they were "singularly free of any form of physical defects . . . one of the strongest and noblest races in the world," and "often reach an extreme and yet healthy old age."[27]

By the mid-1920s, the milk industry, abetted by nutritionists such as McCollum, had persuaded many adult Americans, particularly middle-class ones, to drink milk with their meals, something they had never done before. In 1926, 20 percent of the food budget of the average American "professional" family went for milk, almost double what was spent on meat, and close to four times what was spent for bread and other cereal foods. By 1930 urban Americans were drinking almost twice as much milk as they were in 1916. Consumption continued to rise during the Depression of the 1930s and rose even further during World War II. Whereas coffee had been the beverage of choice among the troops during World War I, in World War II it was milk. In 1945 per capita milk consumption was triple what it had been in 1916.[28] As we shall see, this would by no means be the only time that giant food producers were able

By World War II, the combination of pasteurization laws and clever promotional efforts by the dairy industry had transformed milk from something to be feared into a necessity for the health of adults as well as children. (National Archives)

to use their financial resources and modern marketing techniques to dissipate fears about the healthfulness of their products.

Once Again to the Battlements

Of course, what modern advertising giveth it can also take away, and as the century drew to a close, germophobia reared its head again. The spread of AIDS in the 1980s once more aroused public fears of the spread of infectious diseases. A slew of movies, television programs, and newspaper and magazine articles on bioterrorism stirred this pot. Commercial interests smelled opportunity in this and began stoking germophobia in promoting their soaps, detergents, and even prescription drugs. In 1992 a marketing journal wrote of how in the past four years this promotion of "germ warfare" had led to "amazing" growth in sales of home hygiene products.[29] Then, in 1997 Pfizer introduced a consumer version of Purell, the hand sanitizer it developed for the medical profession, and mounted a fearmongering advertising campaign to promote it. Its success spurred other companies to join in and use germophobia to promote a host of competing disinfectants. Pfizer then responded with Purell 2-Go, small bottles of which could be attached to backpacks, lunch boxes, and key chains, so that people could disinfect their way through the outside world.

The SARS scare in 2004 heightened germophobia even more. Even though SARS was spread by a virus, not a bacterium, most Americans were unaware of the difference. They began snapping up a panoply of germicidal products such as contraptions that sprayed disinfectant on doorknobs, portable subway straps (for those who did not have "City Mitts" anti-microbial gloves), and, to combat germs in airplanes, the Air Supply Ionic Personal Air Purifier. A book called *Germs Are Not for Sharing* instructed children in how to play without touching each other. Those fearing germs on fruits and vegetables could buy the Lotus Sanitizing System, which used an electric charge to infuse tap water with ozone that would kill the bacteria on them—a process not unlike the Electrozone treatment that was purported to turn sea water into a disinfectant a hundred years earlier.[30]

By then, of course, the water supply had once again fallen under suspicion, leading to the enormous growth of the bottled water industry. However, nothing could shake Americans' confidence that breakfast cereals, dairy products, and other foods in brightly colored, tightly wrapped packages were free of harmful bacteria. It all amounted to impressive testimony to the ongoing effectiveness of advertising campaigns such as those mounted by Kellogg's and Nabisco in the early twentieth century. This represented yet another irony: that many of the fears that originated in the industrialization of the food supply were ultimately dissipated by the kind of packaging and marketing that were integral parts of it.

3

Autointoxication and Its Discontents

"Microbes, Our Enemies and Friends": Yogurt to the Rescue

The assault on germs was a two-front war: Not only did it attack bacteria in food before it was eaten; it also assaulted them after the food entered the body. This second front was opened at the turn of nineteenth century, spurred by the discovery that bacteria proliferated at amazing rates in the colon.[1] The scientist leading this charge was the Russian-born bacteriologist Elie Metchnikoff, who warned that deadly bacteria in the colon caused "autointoxication"—a process whereby the body poisoned itself.

Metchnikoff was already a well-established scientist in 1881, when he was forced to abandon his university chair in Odessa by the czarist government's persecution of Jews and liberals (he was both).[2] He moved to the seaside in Sicily, where studying starfish led him to discover that when harmful bacteria enter the bloodstream, they are attacked and destroyed by the blood's white corpuscles, which he called phagocytes, meaning "bacteria-swallowers." This opened a number of important new avenues for combating disease and eventually won him a Nobel Prize in 1908. It also led to the offer of a position at the Pasteur Institute in Paris in 1888 and his selection as Pasteur's successor as director when he died in 1895.

As head of the famous institute, Metchnikoff's pronouncements were often featured in the world's press, illustrated with pictures of him sitting in front of his trusty microscope, with a long gray beard flowing down to his chest, looking every bit like a great scientist. The consequent adulation may have encouraged him to indulge in something not uncommon among highly acclaimed scientists—making pronouncements that were not supported by quite the same rigorous research as that upon which his reputation was based. In particular, there was the claim, which he began making in 1900, that he had "discovered the means of almost indefinitely prolonging life."[3] Over the next seven years, he confidently asserted that not only had he discovered the bacteria that caused the "disease" of old age; he had also found the "helpful and useful" bacteria that would cure it. These life-prolonging bacteria, he said, thrived in a little-known beverage called yogurt.[4]

The key, said Metchnikoff, lay in the human intestines. Evolutionary change had left the human body with significant biological "disharmonies." Our sexual function, for example, is "badly organized," and the process of menstruation cripples and enfeebles women for no good reason. The appendix and wisdom teeth are body parts that may have been useful for survival at an earlier stage but were now superfluous. His *sine qua non* of maladaptation, though, was the large intestine, the enormous tube, often eighteen feet long, whose nether end was the colon. It is, he said, a relic of prehistoric times, when "the need to stop in order to empty the intestines would be a serious disadvantage" for humans chasing or fleeing wild animals. The downside, as it were, of this capacity to delay defecation is that food waste accumulates in the intestines and putrefies there, creating rotting feces that are "an asylum for harmful microbes." Civilization had made this trade-off unnecessary, for cooking performs the original function of the large intestine, which is to make raw food digestible. Yet we remain encumbered by this excessively long organ. It is, he said, a "source of many poisons harmful to the body" that caused "intoxication from within"—what an earlier French scientist had called "autointoxication."[5] If it were not for the maladies caused by "intestinal putrefaction," said Metchnikoff, people would be able to live out their normal life span, which he estimated at 120 to 140 years. "A man who expires at seventy or eighty," he said, "is the victim of accident, cut off in the flower of his days."[6]

The obvious solution to the problem was simply to excise most of the offending organ, and at first Metchnikoff leaned in this direction. He also thought stomachs could be dispensed with and in 1901 wrote that "in the near future it may become the practice of surgery to eliminate the stomach, nearly the whole of the large intestine, and a large part of the small intestine."[7] One of his assistants even proposed that "every child should have its large intestine and appendix surgically removed when two or three years of age."[8]

Some specialists did take this route, albeit by removing only the last section of the large intestine, the colon. Sir William Arbuthnot-Lane, one of England's most prominent surgeons, enthusiastically excised colons from many of his privileged patients, at first to relieve severe constipation, then to cure "autointoxication." Metchnikoff examined three of Arbuthnot-Lane's patients in 1905 and was impressed that their health was "perfect in every way." After being relieved of their terrible constipation, he said, they felt "as though risen from the dead." The fact that nine of the fifty people Arbuthnot-Lane had previously operated on had died as a result of the procedure did raise some eyebrows, but Arbuthnot-Lane claimed that improvements had since been made.[9]

However, although Metchnikoff continued to regard colectomies as effective ways to treat autointoxication, he preferred treating it less invasively, through diet.[10] This was because he had struck upon what seemed like an ingenious way to destroy the nasty bacteria in the colon. He reasoned that since the dangerous bacteria thrived in the colon's alkaline environment, introducing acid-producing bacteria would neutralize its alkalinity and make it inhospitable to the nefarious bacteria. Which acid-forming bacteria to introduce? After doing a rapid survey of reports of long-lived people, he zeroed in on yogurt, the soured milk favored by Bulgarian herdsmen, among whom there were said to be many centenarians. He quickly isolated the bacillus that soured their milk, named it *Bacillus bulgaricus*, and confirmed that it was indeed a prodigious producer of lactic acid—just the thing to neutralize alkalinity in the colon and prevent the noxious bacteria from proliferating there.[11]

It is no surprise, then, that when an English translation of Metchnikoff's book *The Nature of Man* was published in 1903, it caused quite a stir in America. Particularly compelling was his prediction that when people reached their natural life span, they would readily accept death, happy with the knowledge that their natural time had come. Among the converted was William D. Howells, the influential editor of *Harper's Magazine*. He wrote of how impressed he was by Metchnikoff's theory that a proper diet would allow man live to 140 and then see his "instinct for life" diminish and be replaced by an "instinct for death," which would make him welcome the end of his natural span.[12]

The implication that all that was necessary to reach this stage was to drink yogurt was also, needless to say, quite a selling point. A full-page feature on Metchnikoff in the *Washington Times* told of how man could "outwit death" and live for 120 years by drinking "his panacea, sour milk with Bulgarian bacillus." In September 1905 *McClure's Magazine* predicted that "we shall soon have plenty of [yogurt] in America. . . . Professor Metchnikoff takes it daily himself—he keeps a large bowl of it in his laboratory—and with him are many other hard-headed bacteriologists and physicians throughout Europe." A *New York Times* article told of a soda

Elie Metchnikoff, the Nobel Laureate in chemistry who said that dangerous bacteria proliferating in the colon caused "autointoxication," which drastically shortened humans' life span. The popularity of his proposal that eating yogurt would kill these germs and allow people to live to be 140 waned quickly when he died in 1916 at age 71.

counterman offering a customer a glass of "scientifically soured milk," saying, "If you drink that you'll live to be two hundred years old."[13]

Ingesting Metchnikoff's microbes soon became *de rigueur* among the elite. In 1907 a satirical account in the *New York Times* told of a fashionable physician who said there was a "howling mob" surrounding his office, as "the rest of the world and his wife was clamoring for the bacilli of buttermilk." He described a wealthy woman patient who, at a dinner party at her palatial summer home in the Berkshires, served microbes suspended in a gelatin-like drink. She had recently gotten them in Paris from "dear old Professor Metchnikoff, the successor to Pasteur and the great bacteriologist who has . . . postponed the coming of old age well-nigh indefinitely." Metchnikoff had studied the barnyard cow, now "generally famous as the prime circulator of tuberculosis," and concluded that while "ordinary milk contains bad bacilli and you ought to boil it, the little animals in buttermilk are all, or nearly all, benign." The doctor said that the demand for the bacilli was so great that all the pharmacies on Fifth Avenue had run out of them, and "apparently a large percentage of the dyspeptics of the New World were wiring the great scientist for new cultures of his benevolent animals."[14]

In late 1907 Metchnikoff's book *The Prolongation of Life* further popularized his claims that senility was "an autointoxic phenomenon" that was curable by lactic acid. In it, he supported his theory with more reports of the extraordinarily longevity of Bulgarian herdsmen and other yogurt drinkers. "The collection of facts regarding races which live chiefly on soured milk, and among which great ages are common," confirmed, he said, that yogurt prolonged life by stopping the "intestinal putrefaction" that produced the microbes that poisoned the body's tissues and caused "our precocious and unhappy old age."[15]

The "facts" about the longevity of Balkan rustics soon became part of the conventional scientific wisdom. Scientists cited census figures showing that while there was only 1 centenarian for every 100,000 Americans, 1 out of every 2,000 Bulgarians, Rumanians, and Serbs was a centenarian. In 1910 the country's top nutritional scientist, Charles F. Langworthy, head of the Department of Agriculture's nutrition division, said, "Over the whole Balkan Peninsula it is common practice to prepare this yoghurt as a regular everyday article of food, and there is probably no people on earth so healthy as these pastoral Caucasians." The scientist in charge of his department's dairy division said, "There can be little doubt of the definite value of the use of the bacillus Bulgaricus. . . . Yoghurt tried by the experience of the people of the Balkan Peninsula has to its credit the wonderful health and longevity of these hardy races."[16] *The Bacillus of Long Life*, a book by a scientist supporting Metchnikoff's theories, said that yogurt-drinking Bulgarian "peasants" who lived to be 110 or 120 years old were so common that they "fail to excite the wonder of their countrymen."[17]

Yogurt was also credited with specific disease-fighting capabilities. A *New York Times* piece on the longevity of yogurt-drinking "mountain herdsmen and shepherds" said that "tuberculosis is unknown among them and they are free from typhoid fever and all the ills that come from excessive meat-eating." Another article called yogurt "a potent weapon for fighting the germs of certain diseases, such as Bright's disease, rheumatism, typhoid fever, and other ills that devastate humanity." A London producer of Metchnikoff-approved yogurt advertised that it prevented cancer, by killing the cancer-causing bacilli that flourished in the constipated colon, and appendicitis. *The Bacillus of Long Life* claimed mental benefits, saying that yogurt-swilling Bulgarian peasants did not have "the same tendency to mental decay that is so prominent and sad a feature" among "Westerners" in their seventies and eighties.[18] Some months later the *New York Times* reported that two New York City doctors had discovered that *Bacillus bulgaricus* halted the progression of many cases of diabetes and in some cases cured it.[19]

One would expect that Metchnikoff might recoil from these lofty expectations, but if anything he elevated them. In a 1912 article entitled "Why Not Live Forever?" he reported that his recent experiments had indicated that a diet of yogurt and certain other foods (such as ham and eggs) might be able "to transform the entire intestinal flora from a harmful to an innocuous one," thereby eliminating permanently the microbes that cause death. He then polished his Ponce de Leónesque image further by announcing that the telltale sign of aging—gray hair—was preventable. He had discovered that it was caused by germs in the autointoxicated bloodstream that swallowed the cells in the hair that gave it color. Luckily, like other bad bacteria, they were killed by temperatures of 140 degrees Fahrenheit or higher, and he recommended that women preserve their hair color by ironing it, something that those who did so said was successful.[20]

Of course, it did not take long for entrepreneurs to try to cash in on Metchnikoff's fame. Duffy's Malt Whiskey cited Metchnikoff's discovery that "old age is a disease that may be overcome" to support its claim that "constant use" of its product would help men live to be over one hundred. However, it was people claiming to sell products containing Metchnikoff's bacillus who were thickest on the ground. Although yogurt itself could not be imported from the Balkans, it was not too difficult to import the bacillus, or reasonable facsimiles, or to simply pretend that one's product contained them. In England a chocolate maker put the bacillus into bonbons, each of which, it said, "contains some ten millions of the same beneficent bacilli found in Bulgarian sour milk." In America Intesti-Firmin announced that "Metchnikoff's great discovery" was now available in tablets that "contain the finest strains of sour milk ferments of Bulgaria, where people frequently live to be 125 years of age." A dairy in Washington, D.C., sold Sanavita, a "scientifically fermented milk" that was "an application of the discovery made by Metchnikoff of the Pasteur Institute." Soda fountains in New York sold Zoolak, "claimed by Prof. Metchnikoff to be the ELIXIR of LIFE."[21]

It was easiest, though, to sell the remarkable bacillus in the form of "yogurt pills" that could be added to milk. Metchnikoff himself oversaw a process to produce tablets that he said preserved the live bacillus for over three months. To discourage fraud, he cut a deal with the

Parisian manufacturer that allowed only his yogurt to be labeled "Sole Provider of Professor Metchnikoff." Nevertheless, unlicensed imitators were soon cashing in on the Metchnikoff name. Before long at least thirty American companies were promoting pills said to contain *Bacillus bulgaricus*. The *Washington Post*'s medical advice columnist favored this method. He warned that regular milk "becomes putrid and leads to autointoxication" when it reached the intestine, whereas milk soured with yogurt tablets contained "a culture of a friendly germ that antidotes autointoxication."[22]

Such statements implied that any yogurt tablet would do, and this infuriated Metchnikoff. He had licensed only one American company, called "The Metchnikoff Laboratories of America," to produce the pills, and he sued a number of its competitors for using his name. However, such lawsuits were easily evaded. The makers of Intesti-Fermin tablets simply advertised that their product contained "the true culture of Bacillus Bulgaricus, discovered by the scientists of the Pasteur Institute, Paris, and derived from sour ferments of the Bulgarians—many of whom live to be 125 years of age." Another of its ads featured a picture of "the Pasteur Institute, where Professor Metchnikoff discovered the ingredients of Intesti-Fermin." This was accompanied by a drawing of a sturdy-looking "Typical Bulgarian Peasant" and the claim that its ingredients "represent the fruition of a life's work in the part of Professor Metchnikoff." The Lactobacilline company's advertisements offered free copies of a chapter from "PROF. METCHNIKOFF'S . . . 'PROLONGATION OF LIFE.'"[23]

Ultimately, however, none of these companies was able to hit the financial jackpot by making yogurt a significant part of the American diet. Why? One reason was that what I have called "the Golden Age of Food Faddism" was rapidly passing. The impressive array of hucksters making dubious health claims for a host of nostrums—pills, potions, salves, and powders, as well as mechanical contraptions using magnetism, electricity, water, and weights—were now being vigorously denounced by the increasingly powerful practitioners of conventional medicine, whose campaign against fraudulent patent medicines helped spur passage of the Pure Food and Drug Act in 1906. Despite Metchnikoff's prestige, the extraordinary claims for yogurt skirted too close to faddism for them to abide. In 1906, when the American Medical Association set up a Propaganda Department to combat alternative treatments, one of its first actions was to denounce the idea that "autointoxication produced by intestinal obstruction" was a major cause of disease.[24] In 1910 the physician treating a prominent British aristocrat made headlines when he blamed a serious illness she developed on the microbes she ingested in yogurt. Two months later, participants in a special session of the prestigious Royal Society of Medicine denounced the yogurt fad as "a dangerous craze." American doctors now began publicly questioning whether autointoxication really was a danger to health. The oft-cited statistics on the number of centenarians living in the Balkans began to be challenged. In 1916 a popular book derided Metchnikoff as "the modern Ponce de Leon searching for the Fountain of Immortal Youth and finding it in the Milky Whey."[25]

In the end, though, Metchnikoff himself played the major role in undermining his theory. This was largely because he could not resist the temptation to use himself as Exhibit A on its behalf. In 1914, when he was sixty-nine, an interviewer noted that although the great scientist came from "a short-lived family," he was "notoriously vigorous for his years" and, unlike other scientists of his age, showed no signs of slowing down. Metchnikoff attributed this good health and vigor to his diet. "For seventeen years," he said, "I have eaten nothing except what has been cooked: no raw food of any kind, in the form of fruit or otherwise. [He thought raw food brought many more dangerous microbes into the intestine than cooked food.] I find my sugar in dates surrounded with Bulgarian bacilli; the lactic acid comes from the well-known preparation of soured milk."[26] Within a year, though, his heart was failing, and in July 1916 he died, at age seventy-one—a little over halfway through the life span his diet had promised.

Metchnikoff himself did not see his early demise as refuting his theories. As his heart weakened and he realized that the end was near, he speculated that his death was being accelerated by his unusually active life. His actual life experience, he said, was much longer than seventy-one years. He also suspected that there might be some unresolved problem in his colon. He asked a doctor friend to perform a postmortem on him, saying, "Look at the intestines carefully, for I think there is something there now."[27] The rest of the world did not bother to look, for it seemed clear that Metchnikoff's yogurt diet did not work. These doubts were then confirmed by studies showing that yogurt bacilli did not survive any longer than others in the colon. It had also became apparent that Bulgarian herdsmen's life spans were grossly overestimated because fathers, sons, and grandfathers often had identical names, leading census takers to confuse the living with the dead. By 1917 few people would have wagered that yogurt had any future in America.

"Metchnikoff's Mistake"?

Although belief in the curative powers of yogurt suffered a blow with Metchnikoff's death, fear of autointoxication did not go quietly into the night. One reason was its basic appeal to common sense. As James Whorton has shown in his wonderful chronicle of the often-outlandish attempts to combat it, middle-class Americans, especially men, constantly struggled with constipation.[28] Much of this was the related to diet. Before World War I, meat and fried potatoes were the mainstays of middle-class men's diets. Both men and women ate the whitest bread obtainable and disdained whole grain varieties as poor people's fare. Vegetables, when eaten, were usually boiled until mushy, and salads, small and dainty, were dismissed as ladies' food. Most fruits were too seasonal to be of much importance. The upper class had a more varied diet, but this usually meant eating a larger array of expensive meats and seafood, followed by elaborate desserts. Add to that the sedentary lifestyle inaugurated by the new industrial/bureaucratic age, and you have an excellent recipe for digestive pain.[29]

It is no wonder, then, that there were constant complaints about two afflictions: dyspepsia and constipation. Dyspepsia was a catch-all term for the pain and discomfort in the digestive system. Like its female counterpart, neurasthenia, which was thought to be caused by the nerve-racking demands that the new era imposed on well-off women, it had a certain social cachet. Those most susceptible to it were thought to be successful men, whose hard-won material gains had taken a toll on their innards. In October 1874, for instance, the *New York Times* reported matter-of-factly: "Charles H. Derby, a well-known resident of Milford, Mass., hanged himself yesterday. Dyspepsia induced the act."[30]

Constipation was probably the object of even more cures than dyspepsia but, involving as it did the indelicate subject of defecation, it had no social cachet and was mentioned in public only with euphemisms. (The *New York Times* for example, said the colectomy that Arbuthnot-Lane performed on the Duchess of Connaught was aimed at relieving "chronic intestinal obstruction.") Yet, despite this potential for indelicate talk, the rise of the theory of autointoxication in the early 1900s thrust the slow pace at which feces moved through the colon front and center, as it were. It seemed obvious that the longer that disease-causing germs proliferated there, the more likely they were to wreak havoc on the system. Before long, the list of maladies caused by constipation began to rival that ascribed to the housefly. Followers of Arbuthnot-Lane, for example, claimed that removal of the colon could cure cancer, rheumatism, and various nervous disorders.[31] In 1907 a professor at Boston's medical school advised a lawyers' meeting that "autointoxication in certain forms produces insanity," and that "the strange acts of men who drew wills leaving their nurses fortunes might be traced to this cause."[32]

In America Dr. John Harvey Kellogg, head of the famous "Sanitarium" in Battle Creek, Michigan, was able to use the theory of autointoxication to become one of the most respected medical authorities in the nation. The "San" had originated as a Seventh-Day Adventist health resort, and at first Kellogg propounded the ideas of the sect's founder Ellen White. These derived from the doctrines of the early nineteenth-century vegetarian preacher William Sylvester Graham, who said that the sexual emissions stimulated by meat, alcohol, and spicy foods weakened the body and made it prone to disease. When the germ theory of disease cut the ground out from under this, Kellogg turned to Metchnikoff's theory of autointoxication for a new science-based weapon for condemning meat. He now argued that since meat was less easily digested than fruits and vegetables, it must be the major component of the static, putrefying residue in the colon, producing microbes that "flooded" the body with "the most horrible and loathsome poisons."[33] He, too, used yogurt to combat it, although not entirely in the way that Metchnikoff had envisaged. Patients were taken to a special room and given a pint of it, one half of which was consumed orally, the other half of which was delivered into their colons by "enema machines." The latter, said Kellogg, planted "the protective germs where they are most needed and may render most effective service."[34]

Another crusader raising alarms over autointoxication was Horace Fletcher, a wealthy retired businessman whose advocacy of chewing food for extended lengths of time earned him the sobriquet "the Great Masticator." He claimed that "thorough mastication" combated autointoxication by having the body absorb food even before it reached the large intestine, leaving nothing to putrefy there. Kellogg found this persuasive and hung a huge sign saying "Fletcherize" in the Sanatorium's dining hall. He even composed a "Chewing Song" that patients sang before meals as a reminder to do so.[35]

It is important to note that at the time these two characters were by no means regarded as quacks. Kellogg had an MD from Bellevue Hospital in New York and engaged in regular interchanges with world-renowned scientists such as the Russian Ivan Pavlov, soon to be famous for his experiments with salivating dogs. He was also a close associate of the new scientific home economists, college-educated women who were using

An important bastion in the battle against autointoxication: the Laboratory for Fecal Analysis at Dr. John Harvey Kellogg's renowned Sanitarium in Battle Creek, Michigan.

their mastery of the latest findings of nutritional science to consolidate their dominance of teaching about food and health in American schools and colleges. Fletcher was supported by Professor Henry Bowditch of Harvard Medical School, the "Dean of American Physiologists," and was accepted as a collaborator by the prominent Yale chemists Russell Chittenden and Lafayette Mendel, who were in the forefront of nutritional research. They persuaded the chief of staff of the U.S. Army to send a detachment of soldiers to Yale so that Fletcher could test his theories on them. The brilliant Harvard philosopher William James became a "chewer," as did his novelist brother, Henry.[36]

Kellogg and Fletcher did have their differences, including on the important matter of evacuating the toxins. Kellogg said three or four movements a day were necessary, while Fletcher thought that since proper chewing would leave little residue to be swallowed, once every four or five days was best. This would result in dry, odorless feces that were almost pellets, examples of which he sent scientists in the mail. Yet Kellogg was willing to accommodate Fletcher because in practice the Great Masticator's support for drastically lowering protein requirements meant eating much less meat.[37]

Metchnikoff, on the other hand, had remained an unreconstructed carnivore. Moreover, he had warned that the raw fruits and vegetables that Kellogg extolled were laden with microbes and suitable only for the digestive systems of rabbits. Kellogg had consequently distanced himself from the Russian and dropped yogurt as a curative at the "San."[38] When Metchnikoff died in 1916, Kellogg took rather unseemly pleasure in blaming his early demise on his being a flesh-eater. Then he devoted a chapter of his book *Autointoxication, or Intestinal Toxemia*, entitled "Metchnikoff's Mistake" to excoriating the dead scientist, saying that he died because yogurt could not combat the autointoxication caused by his meat-eating. He gleefully recalled one of Metchnikoff's assistants saying to him: "Metchnikoff eats a pound of meat and lets it rot in his colon and then drinks a pint of sour milk to disinfect it. I am not such a fool. I don't eat the meat."[39] That same year, the carnivore Horace Fletcher died at age sixty-eight.

Autointoxication Redux

Kellogg, on the other hand, lived until 1943, when he died at age ninety-one. However, the theory of autointoxication had faded rapidly in the early 1920s, along with the fortunes of the "San." One reason for this was that once Metchnikoff's death snuffed out the yogurt craze, there was little money to be made from selling other products to combat autointoxication. John Harvey Kellogg had refused to join his brother William in commercializing Corn Flakes, which they jointly developed as a vegetarian alternative to the traditional breakfast of meat and eggs. When William went ahead and did so, he did it by promoting the cereal for its convenience and taste, rather than as a cure for autointoxication.[40]

Autointoxication's greatest weakness, though, was that it came to be seen as a threat by the orthodox medical establishment, which was battling to consolidate its dominance of medicine against homeopaths, chiropractors, and a host of other practitioners of what is today called alternative medicine. The essence of creating a profession is to establish a monopoly over expertise, and calls to combat illness through dietary change added nothing to the arsenal of doctors. The few American surgeons performing colectomies to cure autointoxication were not highly regarded by their colleagues. ("To find some of their patients you must ask the sexton," said one skeptical doctor.)[41] Nor did many of them take up colonic irrigation, which required equipping an office with costly waterproof rooms that were difficult to maintain. When the AMA Propaganda Department repudiated the idea that constipation-induced autointoxication was a major cause of disease in 1906, it specifically condemned colonic irrigation as useless and dangerous.[42] Finally, in the early 1920s, scientific studies showing that bacteria in the colon did no harm to the body undermined the entire basis of the autointoxication theory, relegating it, as the AMA hoped, to the realm of quackery.[43]

Yet the idea that bad bacteria in the bowels caused disease proved to have too much commonsensical appeal to disappear completely. Sir Arbuthnot-Lane maintained his high profile in Great Britain, and although he stopped performing colectomies, he did keep pushing the autointoxication envelope. As head of the "New Health Society," he continued to announce such discoveries as that dark-haired people were more susceptible to autointoxication than fair-haired ones. In 1937 he sponsored a speaking tour of England by the influential American health guru Gayelord Hauser, who was reviving fear of autointoxication in the United States.[44] Over the ensuing years, the descendants of Arbuthnot-Lane's aristocratic clientele kept the autointoxication flame flickering long enough for it to emerge later in the century in the social milieu of Princess Diana of Great Britain, who often turned to colonic irrigation as a cure for some of the things that ailed her. Nor did it die in Europe, where in the 1980s and 1990s Chancellor Helmut Kohl and other members of the German elite would go to expensive spas for a week of fasting and irrigation to eliminate the "toxins' from their systems.

By then, fear of autointoxication had also resurfaced in America, as the idea that colonic cleansing to rid the body of toxins became an important element in the burgeoning alternative medicine scene. The number of diseases blamed on autointoxication now surpassed the most extravagant claims of the Metchnikoff era, as newer ones—such as psoriasis, irritable bowel syndrome, Crohn's disease, and chronic fatigue syndrome—were added to the list. This "colon therapy" usually involved antibacterial potions, but on occasion yogurt was employed, as were coffee enemas. Had he lived that long, John Harvey Kellogg would surely have been pleased by this, even though he thought coffee was poisonous.[45]

A Posthumous Victory for Metchnikoff?

Yogurt, of course, made a much more spectacular comeback. During the 1920s and 1930s, the idea that yogurt-eating rustics in the Balkans were exceptionally long-lived never quite disappeared. In a 1926 book called

Outwitting Middle Age, an American doctor hauled out the Bulgarian peasants once again, saying that "out of 7,000,000 people no less than 3,700 are over a hundred and are still going strong. Why? They drink *yogurt*." However, since it was not available fresh in the United States, he could only recommend buying it as tablets in drugstores.[46] When William Hay, the popular reducing diet guru of the late 1930s, recommended yogurt to combat the dangerous bacteria in the digestive system, it was also in the form of pills.[47] His competitor Gayelord Hauser also advised eating fresh yogurt but had to provide a time-consuming recipe for making it at home.

Then, in 1942 Daniel Carasso, a refugee from France, set up a branch of Danone, his family's fresh yogurt business, in Brooklyn, New York, taking on a Swiss-born Spanish Jewish immigrant, Joe Metzger, as a partner. Carasso's father, Isaac, a Sephardic Jew who had moved to Spain from Greece in 1919, had begun producing yogurt in Barcelona from cultures imported from Metchnikoff's old lab at the Pasteur Institute in Paris. He then developed a method for producing it on a large scale, and in 1929 his son Daniel, after whom the firm was named, took this method back to Paris, where he set up a yogurt-making facility. The company did quite well, but the Nazi conquest of France in 1941 forced Daniel to flee to the United States.

The Carasso-Metzger venture was hardly an instant success. Americans had little taste for yogurt, and it was difficult to set up a distribution network for a fresh product that needed refrigeration. They tried Americanizing the name, changing it to Dannon, but until 1947 their market was restricted to a few outlets in the New York City area where the customers were mainly immigrants and a sprinkling of what Metzger's son Juan called "health food fanatics." Then Juan Metzger had the bright idea of appealing to the American sweet tooth by putting a layer of stewed strawberries on the bottom and changing the company slogan from "Doctors recommend it" to "A wonderful snack . . . a delicious dessert."[48]

This boosted sales enormously, but it was the health guru Gayelord Hauser who revived yogurt's reputation as a health food. Tall, handsome, beautifully coiffed, and immaculately dressed, Hauser's charm and good looks were irresistible to women of a certain age. His Road to Damascus story had him dying of "tuberculosis of the hip" at age eighteen, soon after he immigrated to America from Germany.[49] He returned to Europe in search of a cure and encountered an aged practitioner (one version had him as a naturopath in Dresden, another as an old monk in Switzerland, and a third said he was an Austrian immunologist) who cured him by having him drink prodigious amounts of lemon juice and liquefied vegetables. He was soon back in Chicago, promoting herbal potions, the most successful of which was the laxative Swiss Kriss. In 1927 he moved to Hollywood, where he quickly charmed his way into the lives of a number of female stars, including Greta Garbo, by developing special diets to combat sagging skin and aging.[50] Among his recommendations was yogurt, which he said "has long been the staple food of the Bulgarians [who] are noted for their vigor and longevity."

It was not until after the war, though, when Dannon made yogurt widely available commercially, that Hauser, now "Dr. Hauser," made it central to his diets.[51] He deemed it one of the five "wonder foods" in his 1950 best-seller, *Look Younger, Live Longer*, and he proceeded to extol it in a slew of books, newspaper columns, daily radio shows, and weekly television programs.[52] Yogurt sales climbed, as did producers' claims for its healing qualities. Finally, in 1962 the Food and Drug Administration stepped in and issued an order prohibiting them from making health claims for it.[53]

Three years earlier, the conglomerate Beatrice Foods had bought Dannon and began producing and distributing fresh yogurt throughout the nation. The FDA order hardly fazed it, for its marketers had already decided that it was much more profitable to sell it as a sweet snack than as a health food. In 1963 a company spokesman said, "We don't want people to think of yogurt as a product that's good for them. We want them to think of it as something to enjoy eating."[54] The creation of low-fat versions positioned it perfectly for the fat-phobic 1980s, and sales rose impressively. From 1993 to 2004, Americans almost doubled their consumption of fresh yogurt, 85 percent of which was sweetened. Indeed, in 1992 Dannon, which was again in French hands, introduced a yogurt for children so full of sweetened fruit, sugar, and pieces of candy that nutritionists denounced it as a junk food.[55]

Yet despite its sweetness, yogurt still retained a residual reputation for healthfulness.[56] In 2004 the organic yogurt producer Stonyfield Farm listed many of the benefits associated with yogurt. It admitted that these were "folklore," in the sense that they were "widely held beliefs, not proven true," but some of them (it did not say which) "were understood to be quite accurate and accepted." There followed a long list of claims ranging from combating diarrhea to curing colitis, Crohn's disease, and vaginal infections, as well as lowering cholesterol levels.[57]

At first, it seemed that this apparent posthumous triumph for Metchnikoff was a limited one. Despite all the talk about destroying bacteria in the intestines, no one dared claim that yogurt was "life-prolonging." Indeed, even after the government began allowing health claims for processed foods in the late 1990s, most large producers continued to market it as a diet food rather than as a health food. However, in the early 2000s food producers discovered "probiotics." These were varieties of lactobacillus that, although different from Metchnikoff's *Bacillus bulgaricus*, were said to have essentially the same effect: that is, promoting the proliferation of "good bacteria" that killed the bad ones in the colon.[58] Yoplait yogurt now claimed that "Yoptimal immun+" yogurt would "strengthen your immune system with a unique combination of: 2 active probiotic cultures [and] Antioxidants (polyphenols)."[59] Dannon developed two probiotic yogurt drinks whose lactobacilli were said to have opposite effects: Activia combated constipation while DanActive helped cure diarrhea by providing "a positive effect on the balance of the intestinal bacteria."[60]

Probiotics did receive a setback in September 2009, when Dannon was forced to settle a class-action lawsuit for having deceived consumers about their effectiveness. It agreed to reimburse consumers $35 million and modify its claims, although it still said they had scientific backing. But few scientists found these claims convincing. The only claim a panel of scientists that included ones with industry ties could substantiate was that probiotics were effective in combating the kind of diarrhea caused by antibiotics. The claims that came closest to those of Metchnikoff—that probiotics strengthen the immune system—remained unproven. Moreover, although pills seemed to be an effective way of delivering probiotics to the gut, there was no evidence that fresh yogurt was. In September 2009, a review of the claims for yogurt's health benefits by the University of California School of Public Health concluded, "Don't eat yogurt for its health benefits."[61] The next month the European Food Safety Board concluded that not one of the hundreds of "probiotic" strains of bacteria it studied was shown to improve gut health or immunity, and ordered Danone and other yogurt companies to stop claiming that they did.[62] Danone reiterated its confidence that further research would prove their effectiveness, but for the moment, at least, Metchnikoff still awaited vindication. Meanwhile, some months before, in May 2009, Daniel Carasso had died in Paris, at age 103. Unfortunately for the company, unlike Metchnikoff, he never publicly attributed his longevity to yogurt.

4
Bacteria and Beef

That bacteria were everywhere was frightening; that milk could make you sick was disheartening; but charges that bacteria could make eating beef deadly were potentially the most upsetting of all. Thanks in large part to the persistence of the British culinary heritage, beef was by far the preferred meat of nineteen-century Americans, even though they actually consumed much more pork. As a mid-nineteenth-century newspaper proudly proclaimed, "We are essentially a hungry *beef-eating* people."[1] The American food that Mark Twain missed most during his European tour in 1867 was "a mighty porterhouse steak an inch and a half thick, hot and sputtering from the griddle."[2] Twenty years later, the common refrain of new cookbooks was that middle-class cooking was mired in a rut of roast beef, porterhouse steak, and potatoes.[3] It was this iconic status that would help protect beef from some of the next century's most spectacular food scares.

Emerging Unscathed from "the Jungle"

The introduction of refrigerated railroad cars after 1879 was the major factor allowing Americans to indulge in this passion for beef. Before then, scrawny cattle were driven on the hoof into the cities to neighborhood abattoirs, where they were often fed the foul-tasting mash that was a by-product of brewing beer, giving their meat an unpleasant taste.[4] Now cattle from expanding ranches in the West were shipped by train to huge stockyards in midwestern cities like Kansas City and Chicago, where large packers "finished" them on corn and slaughtered them in vast "disassembly" lines. The carcasses were then loaded into refrigerated railroad cars and sent to wholesalers and butchers across the nation.[5]

As a result, fresh beef prices dropped, quality rose, and the number of Americans, particularly middle-class ones, who were able to enjoy fresh beef increased exponentially.[6] At the same time, marked improvements in canning allowed packers to make reasonably priced canned beef available to poorer people. By 1900 Americans were eating much more beef than pork.[7]

But just as this pinnacle was reached, a dark cloud loomed. In December 1898, as much of the nation still gloried in the recent victory over the Spanish in Cuba, General Nelson Miles, the U.S. Army commander there, came out with a shocking charge. He told a commission investigating the war effort that much of the beef supplied to his troops had been adulterated with chemicals. Fresh beef had been treated with "secret" chemicals to hide the fact that it had turned rotten. Canned roast beef was no better: when opened, it smelled as if it had been "embalmed."[8] Another army officer then testified that after his men became sick on fresh beef, he tested it and discovered that it contained the preservative salicylic acid.[9] Subsequently, the great hero of the conflict, Theodore Roosevelt, the new governor of New York, weighed in. He told another army inquiry that the canned beef supplied by the major packers, which he also described as "embalmed," often sickened those of his men who could overcome its distasteful smell and bring themselves to eat it. He himself, he said, would sooner eat his hat.[10]

The implications were obvious: The "Beef Trust"—the five large packers who controlled much of the country's beef supply and had the contracts to supply the army with beef—had used chemicals to mask the deadly bacteria in spoiled meat. The irate mother of a soldier wrote to President McKinley that "thieving corporations will give the boys the worst."[11] However, the government's chief chemist, Harvey Wiley, failed to turn up any chemical additives in the companies' beef, and the two army commissions refused to attribute more than some upset stomachs to the problem, which they blamed on poor handling of the beef in tropical conditions.[12]

Much of the public, however, was not convinced. When the secretary of war made an appearance in Boston, he was heckled by a crowd chanting, "Yah, yah, yah! Beef! Beef!" In the national election campaign of 1900, the Democrats had a field day, charging that the Republican administration had knowingly fed the soldiers embalmed beef.[13] Before long, it was commonly thought that bad beef had killed more American soldiers than had the Spanish.

Confidence in beef was further weakened two years later, when the

HARPER'S WEEKLY

A JOURNAL OF CIVILIZATION

NEW YORK, SATURDAY, AUGUST 13, 1898.

TEN CENTS A COPY.
FOUR DOLLARS A YEAR.

WHO IS THE CRIMINAL?

COLUMBIA TO UNCLE SAM: "THE GOVERNMENT MUST FIX THE RESPONSIBILITY, OR THE COUNTRY WILL."

Columbia asks Uncle Sam, "Who Is the Criminal?" as they stand over poisoned soldiers on ships returning from Cuba after the Spanish-American War in 1898. The culprits were thought to be the "Beef Trust," the meat packers whose "embalmed beef" was said to have killed more soldiers than the Spanish. (*Harper's Weekly*, 1898)

German government banned the import of most American meats. It gave as the reason the very thing that Miles had charged: that borax and other chemicals were used to preserve and camouflage meat that may have been spoiled by bacteria.[14] Then, in 1906 came the publication of Upton Sinclair's sensational novel, *The Jungle*, which described the appalling conditions in the Chicago slaughterhouses that processed much of the nation's beef. The story of how this book led to regulation of the beef industry is well known, but the extraordinary fact that the frightening things it exposed seems to have had hardly any impact on the American love affair with beef has never been recognized.

Sinclair, a socialist, based the book on interviews with slaughterhouse workers and wrote it intending to expose their horrendous working conditions.[15] However, the public was most affected by his disgusting descriptions of the production process itself. Many of slaughtered cattle, he said, were covered with boils or were tubercular, goitered, or dead on arrival at the slaughterhouse. The worst specimens, those covered with boils, went into canned beef. "It was stuff such as this," wrote Sinclair, "that went into the 'embalmed beef' that had killed several times as many United States soldiers as all the bullets of the Spaniards." Much of the rotten meat also went into sausages. Old sausage, moldy and white, that had been rejected by Europeans "would be dosed with borax and glycerine and dumped into the hoppers and made over again for home consumption." The hoppers would also be loaded with "meat that had tumbled out on the floor, in the dirt and sawdust, where workers had tramped and spit uncounted billions of consumption [tuberculosis] germs." In dark storage places, there were large piles of meat covered in rat dung. The packers would put out poisoned bread to get rid of the rats, but the workers would not bother to remove the dead rodents from the pile and "rats, bread, and meat would go into the hoppers together." Fresh beef was hardly safer. When it spoiled, borax was used to camouflage it, just as it had been in Cuba.[16]

Later Sinclair famously said, "I aimed at the public's heart and by accident I hit it in the stomach."[17] However, the disclaimer was rather disingenuous, for he clearly hoped that the gruesome revelations would turn people's stomachs and impel them to rise up against the large packers. He therefore followed up the book with a slew of magazine articles amplifying his disgusting stories. These prompted General Miles to come out in his support, saying the charges were not news to him. Had the matter been taken up seven years before, he claimed, thousands of lives would have been saved. "I believe," he said, "that 3,000 United States soldiers lost their lives because of adulterated, impure, poisonous meat."[18]

The most disgusting of the charges concerned ground meat products. When Theodore Roosevelt, who was now president, let it be known he had been shocked by Sinclair's book, the humorist Peter Finley Dunne had "Mr. Dooley," the Irish immigrant protagonist of his popular newspaper column, imagining the president reading the book over breakfast. Suddenly crying "I'm pizened," he began throwing sausages out the window, one of which exploded and blew the leg off a Secret Service agent and "desthroyed a handsome row iv ol' oak trees."[19]

Whatever Roosevelt's real reaction, he could hardly ignore the loud demands for government action. He invited Sinclair to a much-publicized (sausage-less) lunch at the White House. After hearing the writer out, he threw his weight behind a meat inspection bill that had been languishing in Congress, helping to ensure its quick passage.

Dunne's piece ended by saying, "Since thin, th' President, like th' rest iv us, has become a viggytaryan." However, there is no evidence that Sinclair's revelations caused a significant decline in beef or other meat consumption.[20] There were a number of reasons for this. One is that it was the height of the Progressive Era, a time when the middle classes had considerable confidence that government intervention could cure many of the problems brought on by the new industrial system. Previous scares about the epidemics caused by impure water had led to public health regulations and sanitary improvements that had clearly produced positive results. There was good reason to be confident, then, that the federal Meat Inspection Act of 1906 would clean up the nation's beef supply.

At the outset, this confidence seemed to be justified. As often happened with such reforms, the act, which covered the large packers who shipped meat between states, really owed its quick passage through Congress to the support of the very people who were to be regulated. Initially, these "Big Five" packing companies had resisted the bill, especially the provisions calling for the government to station inspectors in their slaughterhouses. However, faced with the possible loss of confidence in their products sparked by Sinclair's revelations, they soon realized that government inspection would assure consumers of their safety. They threw their support behind the bill, leading Sinclair to charge that Congress had tailored it to their interests.[21] He was right, for meat now left their plants bearing an official U.S. Department of Agriculture stamp, guaranteeing its healthfulness. The packers could now claim, as Armour and Company immediately did, that "the *U.S. Inspection* stamp, on every pound and every package of Armour goods, guarantees purity, wholesomeness, and honest labeling of *all* Armour food products." Better still, it did not cost them a penny, as the many lawmakers who were at their beck and call made sure that the inspection system was funded by taxpayers.[22]

The packers then supplemented this by literally cleaning up their acts, or at least the part of them that could be seen by the public. Barely two

After the stomach-churning descriptions of their slaughterhouses led to the Meat Inspection Act of 1906, the large Chicago meat packers cleaned up their acts and opened parts of their facilities to the public. Visitors were shown reassuring sights such as these women putting "Veribest" labels and U.S. government stamps of approval on Armour products. (Library of Congress)

months after two federal inspectors had been physically sickened when touring its facility, Armour threw it open to the public, welcoming tour groups to gaze upon carefully selected parts of the process. Other slaughterers followed suit, adding a new twist to modern tourism.[23] Another reason the scare had little impact on meat consumption had to do with class. In general, it is the educated middle and upper classes who are most affected by modern food scares. Yet most of the horror stories about beef involved the kinds of canned meats and sausages that were staples of the urban working class, a large proportion of whom were foreign-born immigrants or poorly educated migrants from rural America—hardly the readership of *McClure's* or *The Jungle*.[24] The satirist Dunne's description of the president eating sausage may have been funny, but it was doubtful if such fare was served to Roosevelt, who was about as upper class as one could be in America. Even at breakfast, where they often ate meat, people of his class would eat a small beef steak or smoked ham rather than sausage. The steaks, along with other cuts of beef, would be purchased in butcher shops, where what one historian has called a more or less reassuring relationship of "tension and trust" prevailed between the customer and butcher.[25]

As for the lower classes, suspicion of ground meat products was practically as old as urban life itself, yet it has rarely deterred people from eating them.[26] Moreover, the many immigrants among them would also be reassured by trust in producers of their own ethnicity. And even if they wanted to shun sausages and canned beef and pork, what would they replace them with? Fresh beef, chicken, and pork were generally too expensive for everyday eating.

The relative imperviousness of the working classes to *The Jungle* scare is perhaps best reflected in how little impact it had on the popularity of hamburgers, which in the early twentieth century were regarded as quintessentially working-class food. Butchers were regularly prosecuted for using sulfites to make hamburger meat look fresh, and hygienists warned that eating hamburgers was little better than eating from a garbage pail. Yet their popularity at working-class fairs and carnivals was undiminished. They were also standard fare at the food wagons that clustered outside factory gates.[27]

The ingredients in frankfurters, or hot dogs, which also became working-class favorites in the first two decades of the century, were even more questionable. There were frequent reports in the middle-class press that they were made from a disgusting mixture of ground trimmings, fat, gristle, and corn starch, with saltpeter and red food coloring added to turn the resulting gray sludge pink.[28] A *New York Times* article charged that frankfurters destined for Coney Island were made from offal and scraps from hotels and were "the rottenest of them all." Yet this seems to have had no impact at all on sales at the amusement park's many hot dogs stands.[29] When one particularly gruesome report did cut into sales at the stand owned by Nathan Handwerker (the original Nathan), he quickly revived its fortunes by hiring college students to dress as white-coated physicians and gather there, with stethoscopes around their necks, spreading the rumor that the doctors at the local hospital all ate there.[30]

The years immediately following *The Jungle* scare provided another illustration of how Americans simply regarded beef too highly to give it up. In 1910 many thousands of them roused themselves not over its healthfulness, but over not being able to get enough of it. A mere four years after the publication of Sinclair's book, the very people he looked to for support—workers and reform-minded women—took to the streets not to protest rotten beef, but soaring beef prices. Thousands of members of the nation's women's clubs pledged not to buy beef until its price fell. Five hundred government printing office employees in Washington, D.C., signed on to the boycott, while 4,000 female garment workers in New York City distributed circulars calling on that city's workers to support it. In Pittsburgh, 125,000 men, claiming to represent 600,000 people in their families, signed boycott pledges.[31] But going beef-less proved to be more easily said than done. Too many people, it seemed, agreed with Harvey Wiley, who warned against continuing the boycott, saying that "meat is a necessity, and abstinence from it will result, in the vast majority of cases, in lowered vitality." American men needed to eat plenty of beef, he said, to prevent them from becoming "a race of mollycoddles." The boycott soon petered out, leaving nary a blip on beef consumption.[32]

It all seemed to bear out the German sociologist Werner Sombart's famous observation in 1906 that American socialists' hopes of turning the working class against capitalism were "dashed on the reefs of roast beef and apple pie."[33] Ironically, after the United States entered the Great War in 1917, one of the nation's leading capitalists ran aground on the same reefs. When Herbert Hoover, the head of the government's food conservation program, tried to have Americans cut back on beef consumption by observing "meatless days," his proposal was met with massive resistance from American workers. Rather than declining, beef consumption actually rose 17 percent. The reason, the crestfallen future president concluded, was that the workers were using their newly fattened pay packets to "lay on the porterhouse steak."[34]

Keeping Pandora's Box Closed

Middle-class Americans continued to be wary of ground beef, especially when prepared outside the home, until well into the 1920s. Then, Edgar Ingram, cofounder of the new White Castle chain of hamburger restaurants, was determined to shed the burger's working-class connotations. He set out to attract a middle-class clientele by, in his words, "breaking down the deep-seated prejudice against ground beef." The chain's outlets were clad in tile colored "white for purity." Everything inside was identical and spotless, including the countertops, stools, coffee cups, and cutlery. White-uniformed, well-groomed employees ground the fresh beef and fried the burgers in full view of the customers. Ingram assuaged fears about his burgers' healthfulness by having scientists at the University of Minnesota put a medical student on a diet of nothing but White Castle hamburgers and water for thirteen weeks, eating twenty or more a day for the last few weeks. After the student's health was judged to be unimpaired, Ingram said it proved that someone "could eat nothing but our sandwiches and water, and fully develop all [their] physical and mental faculties." (He never revealed that the student never ate another hamburger again.)[35]

Thanks in large part to White Castle and a slew of imitators, such as White Tower, by the early 1930s hamburgers were well-enough established to withstand another onslaught on beef's safety from the resurgent consumers movement.[36] A 1933 book called *100,000,000 Guinea Pigs*, by Arthur Kallet and Frederick Schlink, two founders of Consumers Union, zeroed in on the dangers of ground beef. It particularly condemned the rampant use of illegal sulfites to camouflage spoiled ground beef. "The hamburger habit," they said, "is about as safe as walking in a garden while the arsenic spray is being applied, and about as safe as getting your meat out of a garbage can standing in the hot sun." In fact, they said, the garbage can was where most butchers' chopped meat belonged. In addition, they charged that government meat inspectors were allowing tubercular cattle to be slaughtered and sold to the public, causing over 5,000 people to die from TB each year.[37]

The book was on the best-seller list for years, and, as we shall see, its critique of chemical additives helped spur some additional oversight in the Food, Drug, and Cosmetic Act of 1938. However, the public's appetite for beef, including ground beef, was unaffected. Hamburgers even became the signature dish at New York City's "21" restaurant, a celebrity favorite known as the city's "haughtiest eatery."[38] During World War II, government propaganda helped raise beef's status even further by emphasizing how the red meat that housewives put on their husbands' plates made them strong and virile. Yet although the supply of beef was adequate, the government appropriated much of the steak for servicemen's messes. The resultant shortage of steak did not sit well with civilians, whose war-fattened pay packets made steak more than affordable, and it became the most sought-after food on the black market.[39]

Beef's unique place in American culture was demonstrated again in 1946, when wartime price controls on beef were abandoned and prices soared by 70 percent. President Truman, hoping to gain popular support, re-imposed the price controls. Cattlemen responded by withholding their animals from market, provoking a shortage. The Republican-dominated press then stirred up a huge public outcry with headlines screaming about a "meat famine." The *New York Times* told of a New York restaurateur who, unable to get enough beef, threw himself off the Brooklyn Bridge. Truman soon caved in and lifted the controls, but his Democratic Party paid a hefty price. It suffered a devastating defeat in the so-called "beefsteak election" of November 1946, when the beef shortage seemed to be on practically every voter's mind.[40]

In the twenty-odd years following the "beefsteak election," hefty rib roasts and thick steaks were an integral part of Americans' self-image as "the best-fed people on earth." A study in the late 1950s showed that beef was by far the most popular main dish for dinner, concluding that "homemakers . . . *think* beef, and apparently *prize beef* above other meats."[41] The suburbs boomed, and houses with backyard barbecues, where husbands in chef's hats grilled steaks and hamburgers, came to symbolize the American Dream. No one thought of questioning the safety of the beef, which was now bought in bright hygienic-looking new suburban supermarkets. Elsewhere in the suburbs, McDonald's was taking the assurances of hygiene pioneered by White Tower to another level. It is no wonder that per capita beef consumption rose by over 70 percent from 1946 to 1966.[42]

In the late 1960s, though, the civil rights movement, opposition to the Vietnam War, startling revelations about hunger in America, and rising concerns about the environment brought a host of critical eyes to bear on American society, including its food. Leading the pack was the consumer advocate Ralph Nader, a lean ascetic who refused to eat any foods that were ground, stuffed, or processed. He began by denouncing the meat industry for turning out "shamburgers" and "fatfurters" that had little meat content. (The latter, he said, were "among America's deadliest missiles.")[43] He then turned to conditions in plants that processed meat for intrastate sale, which were not covered by the 1906 Meat Inspection Act. Consciously modeling his indictment on Upton Sinclair's, he tried to rally support for a bill to bring all packing plants under federal supervision. Echoing the "embalmed beef" charges, he said that, just as in Sinclair's time, dangerous new chemicals were being used to camouflage spoiled beef. Like Sinclair, his most troubling allegations were about ground and processed meats. Network television documentaries explored these allegations and confirmed that packers were using the "4D" animals—dead, dying, diseased, and disabled—for their processed meats.

Unfortunately for Nader, the result also paralleled Sinclair's: The packers quickly realized that federal regulation would help, rather than harm them. They abandoned their opposition to the bill, which shot through Congress in record time. A disappointed Nader said that the packers had managed to put the lid back onto Pandora's box, avoiding further action on a host of other dangers that he had hoped to attack, such as the chemical adulteration of meat, microbiological contamination, misuse of hormones and antibiotics, and pesticide residues.[44]

Nader's pessimism was somewhat misplaced. Eventually, these things did become major issues, but with regard to other foods. Beef, however, remained practically immune from these fears. When beef consumption did decline in the late 1970s and 1980s, it was not because of fears over its safety, but fear of cholesterol.[45]

E. coli in Hamburger Heaven

The greatest test of Americans' attachment to beef, though, began in the 1990s, when repeated reports of frightening deaths and illnesses brought home the fact that the modern beef production industry had suddenly turned the ever-popular hamburger into a random killer. The culprit, which first appeared in 1982, was a dangerous new bacterium, *E. coli* 0157:H7.

The crisis was the direct result of the industrialization of beef production. The practice of "finishing" cattle on corn, rather than the grasses their systems were designed to digest, produced new acids in their stomachs. A new *E. coli* bacterium evolved that was resistant to these acids and made its way into their manure. Since the last months of their lives were spent in immense feedlots, often wallowing knee-deep in manure, the *E. coli* germs were able to migrate back from the manure into their skins and stomachs. Meanwhile, the packinghouse workers' unions had been crushed during the so-called "Reagan Revolution" of the 1980s, leading skilled native-born workers to be replaced by low-paid immigrants with little experience. Their sloppy practices on the killing floors made it possible for the *E. coli* germs to spread from the skins and stomachs into the rest of the animals' carcasses. There they survived and multiplied, wending their way through the various links in the food chain, until reaching humans' stomachs.

Unfortunately, this *E. coli* strain is not only resistant to the acids in cattle's stomachs; it is also resistant to those in ours, and it can kill us if it enters our bloodstream. It is particularly adept at striking the kidneys and is especially dangerous to the young and to the old.[46] To make matters worse, new methods of producing ground beef were almost perfectly designed to spread the bacteria as widely as possible. Beef was being ground in huge centralized facilities where meat from hundreds of carcasses was mixed together, allowing dangerous bacteria from one or two animals to spread exponentially to the rest.

All this hit the headlines in January 1993, when over five hundred people, mainly in the Pacific Northwest, were seriously sickened by the new bacterium. Four of them, all children, died. Most had either eaten at or had contact with people who had eaten at Jack in the Box restaurants, a West Coast fast-food chain. The most frightening aspect of the outbreak was—or should have been—that it highlighted how contagious the new form of *E. coli* was: One of the children who died had not even eaten a hamburger. He had been infected by another child in his play group who had.[47]

The parents of one of the dead boys, hoping that some good might come from his death, joined those demanding that the government begin testing the meat supply for *E. coli*, but despite being granted interviews with President Clinton and members of Congress, they watched in disappointment as nothing was done.[48] This was in large part because the beef industry and the government managed to shift responsibility for avoiding *E. coli* to consumers. The packer who supplied Jack in the Box with the tainted meat had successfully defended itself against the chain's lawsuit by blaming the restaurant for not complying with Washington State regulations ordering that hamburgers be cooked at a high-enough temperature to kill the bacteria. The federal government, unwilling to set up the kind of complex system necessary to test beef for *E. coli* 0157:H7, supported the packer and warned Americans to cook their hamburgers until well-done.[49] When more cases of *E. coli* in ground beef were reported, the USDA responded in a minimal fashion, announcing that it would begin testing a mere 5,000 samples of meat a year from plants and retail stores.[50]

But the news just kept getting worse. Over the next year, the Centers for Disease Control and Prevention recorded sixteen major outbreaks, with twenty-two deaths, mainly among the young and the elderly.[51] In October 1994, after people in three northeastern states were sickened by *E. coli*–infected hamburgers, the USDA finally began random testing of beef. This led to large recalls of beef from Walmart, Safeway, and other chains. Yet the recalls seemed to hardly make a dent in the problem. It was subsequently reported that *E. coli* might still be responsible for as many as 500 deaths and 20,000 illnesses a year.[52]

In 1996 the USDA instituted another new inspection system—one that had the companies themselves check the safety of the meat. Although the system now involved mandatory testing, this was only for salmonella and a benign form of *E. coli*, and not for the killer *E. coli* 0157:H7, which was more expensive to detect. Not surprisingly, there was no diminution in the number of deaths and illnesses caused by eating ground beef.[53]

Recalls soon became a way of life in the beef industry. In August 1997, at the height of backyard hamburger grilling season, the USDA was forced to warn Americans that about 5 million hamburger patties sold across the country may have been tainted with the deadly bacteria. Then it had to order the recall of over 25 million pounds of suspect ground beef produced by two plants in Nebraska.[54] In the summer of 1998, *Time* magazine ran a cover story on "THE KILLER GERM," telling of how ubiquitous and dangerous the bacterium had become. It featured a portrait of a cute little boy from Atlanta who had died in an outbreak alongside a picture of the three-year-old son of Walt Weiss, the Atlanta Braves' star shortstop, who had been stricken in the same outbreak. (The Weiss child was shown enjoying a hot dog while watching his father play in the All-Star Game.)[55]

Yet once again, the scare had no discernible effect on consumption. In 1994, the year after the Jack in the Box tragedy, more hamburgers were ordered in American restaurants than any other dish. That was 5.2 billion of them, or 20.8 hamburgers for every man, woman, and child in the country. On Labor Day of 1997, a week after the massive hamburger recall, a *New York Times* reporter visiting the picnic grounds at Bear Mountain state park found hamburgers sizzling on countless portable grills. When questioned about the scare, most people said they knew of it, but not one said they had refrained from cooking hamburgers because of it. "I like my meat," said a woman cooking them for her church choir. A man who said he was a cook in an Italian restaurant commented, "What are you going to do, you're going to die anyway." A spot check of grocery stores in the area showed no fall-off in beef sales, and one Long Island supermarket had actually run out of ground beef.[56]

This extent of this insouciance is highlighted by comparing it with the response to another ground beef scare, the outbreak of BSE (bovine spongiform encephalopathy), or "mad cow disease," in Great Britain in 1996. It was also the result of changes in cattle "finishing." A new method for making cattle feed from the bones and leftovers from abattoirs failed to destroy dangerous microbes in some of the diseased animals that were ground into meal. As with *E. coli* 0157:H7, the disease jumped from cattle to humans, in this case causing a terrifying brain-destroying affliction called variant Creutzfeldt-Jakob disease (vCJD). As with *E. coli*, it spread most easily in ground beef. When the British government finally acknowledged what was happening in March 1996, consumers around the world panicked. Beef consumption in Great Britain, Germany, and Italy immediately dropped by 40 percent. In Greece it dropped more. Even in France, where labels on most beef in France guaranteed that it did not originate from diseased herds, consumption declined by 20 percent. Throughout the world—including in far-off nations such as China, Japan, and Korea, where no cattle or people had been stricken—people stopped eating beef.[57] Yet in 1995 and 1996, the total number of deaths in the epicenter of the panic, Great Britain, where the vast majority of cattle with BSE were found, was thirteen, about one-thirtieth the number of Americans who were estimated to die each year from *E. coli*.[58] However, the only people unaffected by the panic seemed to be Americans, who, even at a time when mistrust of government was rampant, put their faith in government assurances that no American cattle suffered from BSE.

In late 1997, a solution to the *E. coli* problem seemed at hand when the Food and Drug Administration approved radiating meat as a perfectly safe way of killing the bacteria that had no effect on the taste of the meat.[59] However, unreasonable public fears that irradiating food would make it radioactive could not be overcome, and the project was quietly shelved. Another system for inspecting the plants was phased in from 1998 to 2000, but it hardly made a dent in the problem. The announcements of outbreaks and recalls, followed by halfhearted government assurances that the system was working, continued with alarming regularity.[60]

The time seemed ripe, then, for heirs of Upton Sinclair to again come to the fore, and they tried. In 2001 Eric Schlosser's book *Fast Food Nation* followed the foods used in the fast-food industry from farm to plate—what he called "the dark side of the American meal." Like Sinclair, Schlosser highlighted how the abysmal working conditions of the low-paid immigrants in the slaughterhouses contributed to the spread of *E. coli* 0157:H7 and other food-borne diseases. The Jack in the Box outbreak enabled Schlosser to go further than Sinclair and link specific deaths to these practices, providing heartrending descriptions of how children, who are most susceptible to the bacteria, died.[61]

Like Sinclair, Schlosser called for government intervention to remedy the situation, but the political map was now much different. For one thing, the meat packers now wielded even more political power than they had in 1906. Back then, the Big Five dominated only the interstate market for beef. Now four companies slaughtered more than 85 percent of the entire nation's beef.[62] In 1906 Progressives seeking tighter control over industry were powerful voices in the White House and Congress. Now the conservatives who dominated Washington sought to loosen government controls over business, not tighten them. In addition, the obvious solution for the *E. coli* problem, the radiation process recommended by the FDA, was anathema to the kind of middle-class people who were disturbed by Schlosser's revelations.[63] They wanted less processing of their foods, not more, and found the idea of "nuking" them terrifying. Yet there was no other legislation waiting in the wings, such as the meat inspection bill of Sinclair's time, promising to solve the problem.

Perhaps the absence of any obvious governmental solution explains why Michael Pollan, author of the 2006 best-seller *The Omnivore's Dilemma*, leaned toward individual, rather than governmental solutions to the problem. His book graphically describes the manure-filled feedlots and sloppy slaughtering practices that practically guaranteed that *E. coli* would spread from cattle to humans. Yet in it and his follow-up book, *In Defense of Food*, Pollan's main proposal is for Americans to eat much less beef, and when they do, to seek out grass-fed beef, since it is much less likely to harbor the dangerous strain of *E. coli*.[64]

The minor extent to which the gut-churning descriptions of where beef came from affected its continuing popularity was highlighted in late September 2006, when *E. coli* originating in manure from a large cattle-raising operation caused three people to die and hundreds to be sickened in twenty-six states. This time, however, the fecal matter had seeped into a neighboring large corporate farm, infecting baby spinach. As with the beef outbreaks, new production methods practically guaranteed that the *E. coli* would spread on a previously unimaginable scale: the company mixed its farm's greens with those from a number of other farms into 26 million servings of greens and salad every week, including some that was labeled "organic." Within days the Food and Drug Administration issuing a warning against eating the company's bagged baby spinach and salads and ordered all of its produce to be recalled. The FDA then assured consumers that the source of the infection had been found, that the problem was corrected, and that any suspect spinach had been recalled. Despite this, consumption of fresh spinach plummeted by over 60 percent and subsequently recovered very, very slowly.[65] Indeed, sales were so damaged that the corporate vegetable growers were forced to swallow their historic opposition to any kind of regulation and ask the state to impose mandatory food safety guidelines, enforced by state inspectors. Government regulation, said their trade association's president, was "essential to restore public confidence."[66]

Meanwhile, beef easily weathered an onslaught of equally frightening news. In late 2006 *E. coli* from tainted ground beef sickened almost one hundred customers at Taco Bell restaurants in the Northeast, yet it caused a mere 5 percent drop in sales, from which the company quickly recovered.[67] Faith in government oversight could hardly have played a role in this. In late 2007, after frozen hamburgers produced by a New Jersey company caused illness and kidney failure among consumers in the East and Midwest, the USDA ordered a recall of 21.7 million pounds of its product. There then followed the usual assurances from a USDA official that "the American meat supply is the safest in the world.... A recall like this does show that we are on the job, we are doing our inspections, our investigation, and we respond when we find problems to make sure that supply is safe."[68] Some days later it was revealed that the department had not been "on the job" at all—that it had sat on the evidence of the outbreak for eighteen days, allowing people to buy and eat many thousands of possibly tainted burgers. An official then admitted, "There is room for improvement."[69]

Indeed there was. In 2007 *E. coli* contamination led to twenty-one recalls of beef, compared to eight in 2006 and five in 2005. Yet much still evaded detection. In 2007 more than 25 million pounds of tainted beef were said to have gone on the market in America. This was dwarfed in

February 2008, when the USDA—after being alerted not by its own inspectors but by the Humane Society—was forced to recall what remained of 143 million pounds of beef from a California feedlot that had entered the food supply since February 2006. The operator was videotaped processing cows, called "downers," that were too sick to walk. These animals had an elevated risk of carrying *E. coli* because they spent the last days of their lives down on the ground, mired in feces, exposing most of their skin to the bacteria. Thirty-seven million tons of the company's ground beef had ended up as burgers, tacos, and chili in school lunch programs. Practically every major food conglomerate had used it: General Mills, Nestlé, ConAgra, and Heinz, which was forced to recall 40,000 cases of Boston Market Lasagna with Meat Sauce.[70]

Meanwhile, the government implicitly confirmed the critics' charges that the system of giant feedlots and slaughterhouses made *E. coli* too ubiquitous and too elusive to be entirely eliminated from the beef supply. The USDA and FDA now seemed to see their major role in combating it as issuing recalls of *E. coli*-contaminated beef after consumers had sickened or died. The USDA continued to advise consumers to ensure that their hamburgers were well-done, but since one in four burgers that are brown in the middle are not hot enough to kill dangerous bacteria, the only way for wary diners to ensure the safety of their burgers would seem to be to go into restaurant kitchens with meat thermometers to test burgers as soon as they come off the griddle.[71] Another solution to the hamburger problem ended up echoing the embalmed beef controversy. In 2002 an inventor named Eldon Roth gained government approval for a method that allowed him to make ground beef out of the discarded fatty trimmings from slaughterhouses that had previously been used for pet food and oil. His company, Beef Products, used centrifuges to separate the protein residues from the fat. The resulting ground beef-like substance (one microbiologist described it as "pink slime") was then treated with ammonia gas, which killed any pathogens for *E. coli* 0157:H7 and salmonella.[72] After being frozen into sixty-pound blocks, it was sold to school lunch programs, prisons, Cargill, McDonald's, Burger King, and many other retailers looking for ways to reduce their beef costs. The problem was that, just as in Theodore Roosevelt's time, some people using the product recoiled from the pungent smell of the ammonia. It was then alleged that, as a result of these complaints, the company sometimes cut the amount of ammonia to levels below that at which it would kill the pathogens. In 2010 the discovery of the two pathogens in batches of its meat caused school lunch officials to suspend their purchases and cast a cloud over the process. No one said, "I'd rather eat my hat," but many people were thinking similar thoughts.[73]

Herein lay a key to the continuing insouciance of most Americans regarding scares over beef: they rarely involve, as did the use of ammonia during the "embalmed beef" scandal, that powerful force in shaping people's food habits—disgust. Instead, the warnings rely on the central message of twentieth-century nutritional science, that good taste is not a guide to the healthfulness of food. To most Americans, beef tasted too good to pay much heed to that.

With radiation too hot a potato to handle and ways of "embalming" beef again in disrepute, demands for more intense government regulation of slaughtering again resounded in Washington, D.C. But a simple solution like the Meat Inspection Act of 1906 was no longer possible, for the *E. coli* problem seemed intractably connected with the entire system of giant feedlots and slaughterhouses. Some suggested abolishing the system and returning to a beef supply that was primarily grass fed. But even if this were politically possible, which it is not, it would make high-quality fresh beef what it was in the mid-nineteenth century: an expensive food for the well-off. Some critics admitted this and advocated individual solutions—that people buy pricey grass-fed beef but eat much less of it.[74] However, given Americans' historic attachment to beef, there was little prospect of this becoming a mass movement.

On the other hand, in 2010, after spinach, peanut butter, and eggs were involved in nasty cases of food poisoning, consumer advocates and industry representatives managed to gain rare bipartisan support in Congress for a bill, which passed in January 2011, giving the FDA new powers to order recalls and supervise food production. Significantly, the only industries that escaped this heightened surveillance were the beef and other meat industries, which remained under the benign gaze of the Department of Agriculture.[75]

5

Lucrezia Borgias in the Kitchen?

It is hard to think of a more frightening word in the food lexicon than "poison." As I have noted, our omnivorous hunter-gatherer ancestors had to be constantly on the alert for poisonous foods. The agricultural revolution allowed humans to grow foods they knew were safe, but the market economy that accompanied it brought new worries: unscrupulous middlemen could increase their profits by adulterating food with dangerous substances. The new ways of producing, preserving, and transporting foods that arose in the nineteenth century heightened these fears by widening the gap between those who produced foods and those who consumed them. The spectacular growth of cities in the late nineteenth and early twentieth century turned this gap into a chasm.

For the most part, the chicanery involved non-lethal tricks, such as using chalk to whiten milk and bread. However, the development of new chemical preservatives in the later nineteenth century aroused fears that not only could they mask spoiled, possibly poisonous foods, but that they were themselves poisonous. By 1895 twenty-seven states and a number of cities had already passed laws against "adulterating" food with them.[1]

The state and municipal laws against adulteration produced a crazy-quilt of regulations, administered by officials with varying degrees of expertise. Exposés of the ineffectiveness of these laws led to demands that the federal government step in and regulate the new additives. This outcry merged with scares about dangerous patent medicines to help secure the passage of the federal Pure Food and Drug Act, which became law on June 30, 1906, the same day as the Meat Inspection Act.[2] This successful campaign is usually credited with restoring Americans' confidence in the healthfulness of their food. However, it could be said that quite the reverse was the case: that the slew of fearsome stories about poisonous food additives actually sowed the seeds for the fears of processed foods that would periodically sweep the nation in the years to follow.

Harvey Wiley and His "Poison Squads"

The person most responsible for passage of the Food and Drug Act was its primary author, Harvey W. Wiley. The son of evangelical Protestants whose Indiana farmhouse resonated with Reverend Sylvester Graham's denunciations of processed foods, Wiley was raised with the idea that there was something immoral about how "pure food" was denatured once it left the farm gates.[3] After working his way through college, he became a professor of chemistry at newly founded Purdue University. Two years later, in 1878, he made a pilgrimage to Germany, the mecca of research in chemistry, where he mastered new techniques for analyzing the composition of foods. Upon his return, he quickly made a name for himself by exposing fraudulent foods sold in Indiana. In 1883, after being turned down for the university presidency, he moved to Washington, D.C., as chief of the U.S. Department of Agriculture's Bureau of Chemistry.[4]

As soon as he got to Washington, Wiley set up a laboratory that began issuing reports condemning the adulterants that were being used to alter practically every food Americans ate. He discovered pepper that contained charcoal and ground coffee bulked up with chicory, acorns, and seeds. He was particularly outraged by a "strawberry jam" that consisted of nothing more than glucose, artificial coloring, "ethereal salt," and hay seeds that imitated the berry seeds.[5] Such chicanery, he said, was an affront to God, who had provided such natural bounty to America.[6]

However, while Wiley was relentless in tracking down adulteration, he was reluctant to question the safety of the new chemical preservatives that food processors were beginning to use, perhaps because many of them were concocted by the same German chemists at whose feet he had sat. When questions were raised about the health risks of boric acid, benzoic acid, salicylic acid, and formaldehyde, as well as new artificial food colorings derived from coal tar, he suggested only that they be listed on the products' labels, so that consumers could decide for themselves whether they wished to use them. Moreover, as we have seen, he refused to jump on the anti-preservative bandwagon during the "embalmed beef" controversy.[7]

However, that scandal proved to be a turning point. The *New York Times*

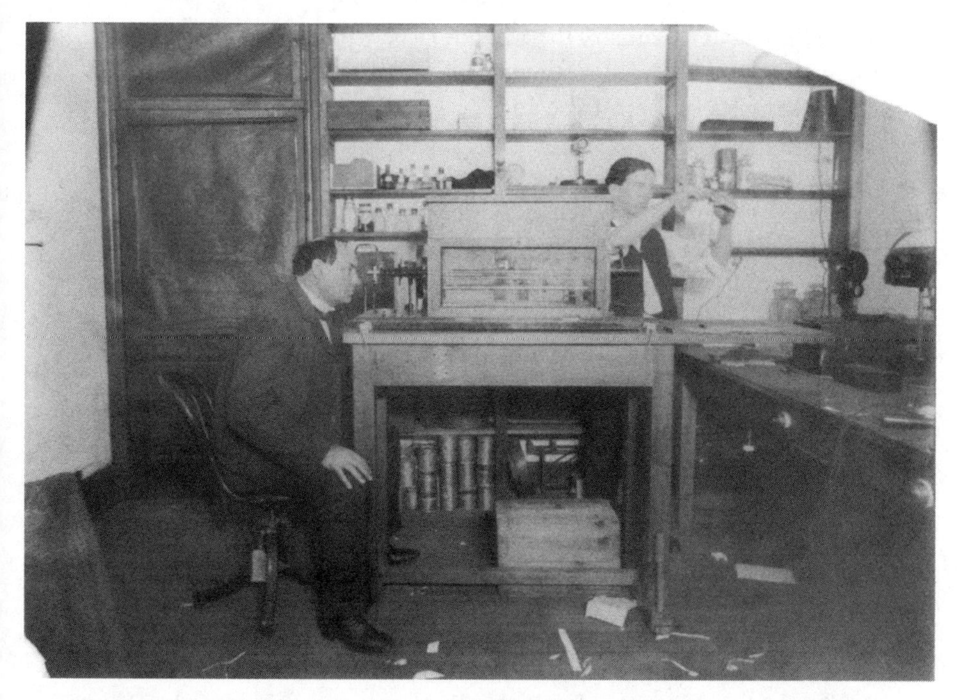

Harvey W. Wiley, chief of the federal Bureau of Chemistry, in his agency's laboratory, searching out poisonous additives in food. His high-profile investigations were aimed at reassuring the public that the government was protecting them, but they also helped stoked suspicions of what food processors were doing to food. (Library of Congress)

said that "since the 'embalmed' beef scandal . . . every one has looked with doubt on any food which is known to contain preservatives."[8] "Muckraking" journalists began exposing how unscrupulous food producers were using allegedly dangerous chemicals to preserve foods and resorting to subterfuges such as camouflaging rotten eggs with chemical deodorants.[9] The idea that greedy businessmen were poisoning foods proved particularly attractive to Progressive reformers, who blamed giant corporations for poisoning American politics. In 1902, after the powerful General Federation of Women's Clubs and the National Consumers League joined in the growing calls for federal government action, Wiley performed a deft turnabout.[10] He persuaded Congress to appropriate $5,000 for him to recruit a group of twelve young male volunteers for a test of the suspect chemicals' impact on their health. In the first test, the brave young fellows—low-paid government clerks happy to have three free meals a day—assembled three times a day in a special dining room at the Bureau of Chemistry to eat foods laced with borax and boric acid, the prime suspects in "embalmed beef." After their dinner, they would collect samples of their urine and feces and leave them with Wiley for laboratory analysis.[11]

Few people would likely have known of the daring young men had not a reporter for the *Washington Post* dubbed them the "Poison Squad." At first, Wiley was taken aback by this and refused to use the word "poison." But it struck a particular chord with the public, since scientists searching for the causes of disease had recently been using themselves as guinea pigs, sometimes with famously fatal results. The *New York Times* told of how Wiley "cooked the doctored food and observed the slow approach towards death of his Poison Squad, measuring their daily progress in the interest of science." It reported that although Wiley would not reveal what they were eating, "it is known that each of the martyrs to science ate several ounces of poison—about the same amount fed to soldiers in Cuba in the unpleasantness with Spain."[12]

Wiley's reluctance to use the term "Poison Squad" faded quickly in the face of the ensuing wave of publicity. The press dubbed him with the colorful, if ambiguous, sobriquet "Old Borax." Songs about the Poison Squad swept through the vaudeville circuit. One of them said, "They take a batch of poison every time they eat a meal. For breakfast they get

One of Wiley's "Poison Squads"—young government clerks recruited to test the dangers of food additives—in their special dining room. Although none of them was ever really poisoned, the alarming name brought national attention to Wiley's campaign for a "pure food" law.

cyanide of liver, coffin shaped." Another concluded, "Next week he'll give them mothballs, a la Newburgh or else plain; O, they may get over it but they'll never look the same." A popular poem about them said, "We're on the hunt for a toxic dope that's certain to kill, sans fail." Much of the merriment at the Washington press corps' annual Gridiron Dinner, which made fun of President Theodore Roosevelt and other luminaries, revolved around a fake "Poison Squad."[13] Three months into the test, the *New York Times* reported, "Few scientific experiments have attracted more attention from the general public than the chemical boarding house of the Poison Squad."[14]

As with *The Jungle*, contemporaries thought that the tales of the Poison Squad's travails were the crucial element in rallying public support for legislation that had been stalled in Congress—in this case, a national pure food and drug bill. One of the bill's advocates later wrote, "The picture of that little 'Poison Squad' in Washington swallowing its daily doses of borax caught first the fancy of the press and then that of the public. In a few months a single sensational venture did what twenty-three preceding years of laborious toil had failed to accomplish."[15]

However, like the meat inspection bill, the support of the very interests that were to be regulated—in this case, food processors—was also crucial.[16] In late 1903 Wiley gained the support of one of their trade associations by arguing that, administered by a (presumably sympathetic) specialist such as he, the bill would free them from the maze of often-contradictory state and municipal regulations.[17] His office then produced reports revealing how small processors were able to undercut the large firms' prices by using chemicals to camouflage inferior or rotting products.[18] This brought the canning giant Henry J. Heinz on board. He arranged for Wiley to meet with the other large canners and explain to them how the act would protect them from this kind of competition. Their trade association then threw its support behind the bill, saying that Wiley's bureau would be able to use the data gleaned from his "poison boarding house" experiments to help them.[19]

Wiley later said that without Heinz's help, he "would have lost the battle for pure food," and this gratitude was reflected in Wiley's final draft of the bill.[20] Only one item, catsup, would be exempt from the proposed restrictions on chemical preservatives. Wiley said he was "lenient" with it because without benzoate of soda, it would spoil as soon as a bottle was opened, and, anyway, people consumed it in very small quantities. It also happened to be Heinz's cash cow.[21]

In the first years following passage of the act, the large processors' support seemed to pay off. The bureau used the act's requirement that any additives to foods be listed on their labels to expose numerous cases of adulteration. Two years after the act's passage, Wiley boasted that he had initiated 176 criminal prosecutions for misbranding and misrepresentation. Most had to do with fakery in the sleazy patent medicine industry, but a considerable number involved mislabeling food. None of the miscreants, though, were big fish. Instead, they involved such cases as Vermont maple sugar vendors using cane sugar in their product, and a New Orleans sugar distributor selling sugar made from corn syrup as cane sugar.[22] In June 1911 Wiley took credit for having initiated 804 food prosecutions over the past three years. Yet the vast majority were for adulteration and misrepresentation by small-timers. (Over one hundred of them were against people caught watering milk.) H. J. Heinz must have been particularly pleased that Wiley's department seized and destroyed large quantities of catsup made by smaller producers, declaring the products to be "filthy, putrid, made of decomposed vegetable matter, to contain chemical preservatives, to be filled with bacteria, and to be utterly unfit for food purposes of any kind."[23]

The stream of headlines about the bureau's prosecutions for adulteration helped maintain public suspicions that there was still widespread tampering with their food, but at least they provided some reassurance that the government was protecting them. Poisonous additives, however, were another story. The act banned the sale of foods containing "any added poisonous or other added deleterious ingredient which may render such article injurious to health."[24] To Wiley, its intent was clear: the "pure" in the Pure Food Act, he said, meant food that was free of "poisons"—chemical additives that were either injurious in themselves or ones that masked foods that had deteriorated and become "poisonous."[25] (He downplayed the competing threat—germs—saying that the dangers of bacteria were being "overplayed" by scientists and that they were not half as dangerous as people had been led to believe.)[26] Yet he found it impossible to nail these chemicals to the wall. As a result, the public was repeatedly exposed to warnings about poisons in food, yet was never assured that the problems had been resolved.

Wiley's problem was that although the term "Poison Squad" conjured up visions of death, or at least serious illness, five years of feeding successive Poison Squads a number of chemical additives failed to produce any evidence that any of them were what is generally considered "poisonous." He therefore tried to put the worst possible face on them. His long 1904 report on borax and boric acid concluded that if consumed in small doses over long periods or large doses in a short time, they "create disturbances of appetite, digestion, and of health." Large doses could also cause "a slight clouding of the mental processes" and nausea.[27] His second set of reports, issued in 1906, condemned sulfurous acid, benzoic acid, and formaldehyde, but, as in his first report, nowhere was there evidence that the chemicals caused serious illness, let alone the Holy Grail of food scares—death. Indeed, when an ex-member of the Poison Squad died of tuberculosis and the young man's mother called him "a martyr to science," Wiley dismissed as "absurd" the idea that his death was connected to all the borax the poor young man had consumed.[28] As for salicylic acid, Wiley had to admit that although large quantities of it did seem to at first stimulate and then depress his subjects' appetites, it was not nearly as harmful "as has been generally supposed."[29]

Part of Wiley's problem was that he thought that if a chemical prevented or slowed down spoilage in food, it must also harm the digestive system. In fact, this was not at all the case.[30] Moreover, his experiments were deeply flawed, suffering from defects that would characterize many subsequent studies of human nutrition: There were too few subjects; there was no way of ensuring that they were not consuming other things when they went home in the evening (one young clerk eventually confessed to eating and drinking at parties); there was no control group of people who did not consume the additives; there were too many variables in terms of their personal health histories (one young man, for example, was suffering from malaria, nephritis, and other serious ailments); and there was no follow-up study of long-term effects. Wiley made much of how a number of the men consuming benzoate of soda developed colds, fever, headaches, nausea, and sore throats, but neglected to mention that during that time Washington was hit with an epidemic of influenza, whose symptoms were essentially the same as those he ascribed to the preservative.[31] In the end, he was forced to argue that although it seemed that the chemicals did no harm in the small doses in which they were used in food, there was no proof that these small amounts would not prove dangerous when taken over many years.[32]

None of this deterred Wiley from repeatedly warning the public about the dangers of chemical additives. In June 1907 his Bureau of Chemistry issued its first ruling on preservatives. It approved traditional methods—such as salting, smoking, sweetening, and pickling—but said it would ban the additives that had gained notoriety during the "embalmed beef" controversy: borax, boric acid, and salicylic acid.[33] Sulfur dioxide and benzoate of soda were given temporary reprieves, but Wiley made it clear that he would soon seek to ban them and all other chemical preservatives. Testifying before the House Agriculture Committee in January 1908, he said that since "drugs" such as borax, benzoic acid, benzoate of soda, sulfate of copper, sulfur dioxide, formaldehyde, and salicylic acid were expelled from the body by the kidneys, their use as preservatives was almost certainly a cause of kidney disease, which was "so prevalent among Americans."[34] In a magazine interview, he went further: Preservatives, he said, were the cause of "universal dyspepsia and . . . an enormous increase of kidney diseases. All the preservatives have cumulative effect, and all of them attack the kidneys—that is, all of them except copper sulphate, which commonly kills before it has time to get to the kidneys."[35]

Wiley hoped that the large food processors would see that banning these additives would counteract the widespread suspicions of processed foods that were now in the air, and some did.[36] Sebastian Mueller, a prominent chemist who was a consultant for large processors, said that such a ban would combat the "deplorable prejudice against all American food products" that had "inflamed peoples' minds" as a result of "the late Chicago meat agitation" and "restore confidence in all prepared foods."[37]

However, many processors were too dependent on chemical preservatives to indulge in such long-range thinking. Some battled the bans head-on, arguing that if instead of ingesting chemicals, Wiley's young subjects had consumed large amounts of the vinegar, salt, smoke, and the other traditional preservatives that Wiley approved, they would have been "*really* sick."[38] Others had politicians intervene to gain exemptions for them. Henry Cabot Lodge, the powerful senator from Massachusetts, had salt cod exempted from the prohibition against the use of borax. Molasses makers got southern state health officials to support their refusal to stop using sulfur, which had been used for over one hundred years to clarify molasses and prevent it from separating. California dried-fruit processors persuaded the government to turn a blind eye to their continued use of sulfur as a preservative.[39]

Although these lobbies ate away at his power in Washington, Wiley's repeated warnings about poisons in food helped solidify his reputation as someone protecting the public from chicanery. He dug his political and professional grave, though, by backing the wrong horse—the chemical preservative benzoate of soda—in the Worst Poisons Stakes. A 1904 Poison Squad experiment had convinced Wiley that it was dangerous, but, as we have seen, probably because H. J. Heinz needed it in his catsup, he had allowed processors to use it if they listed it on their labels. However, in late 1906 Heinz's company discovered a secret process that allowed it to make what it called "ketchup" without benzoate of soda , and in early 1907 Wiley proposed a complete ban, to take effect the next year. Without the additive, Heinz's competitors' catsups would spoil within days after being opened.[40]

In January 1908 President Roosevelt called Wiley to the White House to meet with him and a number of Republican food processors who were complaining about the damage his proposed bans on benzoate of soda and saccharin would do to them. According to Wiley, the president asked the secretary of agriculture if adding benzoate of soda to food was injurious to health, to which the secretary replied that he thought it was. The president then asked the same question of Wiley, who replied, "I do not think; I know. I have tried it on healthy young men and it has made them ill." Roosevelt then turned to the businessmen, struck the table with a ringing blow, and declared, "Gentlemen, if this drug is injurious you shall not put it in food." According to Wiley, the matter would have stopped there had not one of the businessmen then brought up the question of saccharin. Wiley's proposed ban, he said, would cost his canning firm thousands of dollars it was saving by using it instead of sugar in its canned sweet corn. Wiley then intervened, saying that this deceived consumers and that "moreover health is being threatened by this drug." The president, suddenly enraged, turned on Wiley and, "with a fierce visage," said, "Anyone who says saccharin is injurious is an idiot. Dr. Rixey gives it to me every day."[41]

Two days later Roosevelt appointed a board of scientific experts, headed by Ira Remsen, the prominent chemist who was president of the Johns Hopkins University, to decide on the dangers posed by preservatives. Acceding to the president's wishes, it started with benzoate of soda and saccharin. One of its members, the Yale professor Russell H. Chittenden, fed benzoate of soda to a "Poison Squad" of six graduate students for much longer than had Wiley. Two other board members also subjected their students to substantial doses of the chemical. In January 1909 the Remsen Board announced that no harm had come to any of the students, even from high doses, and that benzoate of soda, used in the relatively small proportions allowed by the government, was not at all harmful.[42]

Some of Wiley's friends were sure he would now resign, but not only did he cling to his post; he counterattacked. He rallied Heinz and a number of other large canners into a wholesale attack on the Remsen Board, involving the press, trade associations, and advertising. The "muckraking" journalists Samuel Hopkins Adams and Mark Sullivan, whose exposés of additives had rallied public support for passage of the Pure Food and Drug Act, joined in. They condemned the Remsen Board for having gutted the act, with the connivance of the president and the secretary of agriculture, and turning a blind eye to poisons in the food supply. A movement of unknown origin arose to nominate Wiley for the Nobel Prize in chemistry.[43]

In August 1909 Wiley went to the convention of the heads of the state food and dairy departments, which the year before had condemned benzoate of soda. He was now confident that they would join him in denouncing the Remsen Board and support a new investigation of benzoate of soda. While waiting for the vote, he warned that

> modern housewives are veritable Lucrezia Borgias, handing out poison from the ice box, from the broiler and the skillet, and the little tins of dinner she buys when breathlessly rushing home after her exciting bridge games at the club. . . . The average ice box is a charnel house, which not only holds death, but spreads it.[44]

Much to his disappointment, though, the conference supported the Remsen Board's findings. A month later the American Institute of Homeopathy, which claimed some 25,000 members, also reversed itself and declared there was no convincing proof of Wiley's charges.[45]

Again, Wiley was expected to resign from the government, and again he refused. Responding to a deluge of telegrams from women's clubs and consumers organizations pleading with him to stay on, he said, "I will remain where I am, and continue the fight until they force me out." Nor would he concede an inch on benzoate, even though practically every university chemist in the country now seemed to disagree with him. Instead, he denounced the scientists who disputed his claims as lackeys in the pay of the food industry.[46]

Wiley tried to rally public support with attacks on the safety of other additives. In 1909 he went after alum, which was used to turn flaccid cucumbers into stiff pickles. "What benzoate of soda is to the decaying tomato and borax is to embalmed beef[,] alum is to the limp and lifeless cucumber," he declared. "The public does not appreciate the woes and misery concealed beneath the verdant jacket of the innocent-looking pickle." What were these woes? Some students at a Philadelphia girls' school had suffered upset stomachs and nausea after eating pickles. A toxicologist who supported Wiley admitted that one would have to ingest huge quantities of pickles for the alum in it to cause indigestion. However, he said, "we know that neurotic and cholorotic females often take large quantities of pickles, and for them it might be a very serious matter."[47]

He also charged that caffeine was a harmful and addictive chemical. It and cocaine, he said, were the most commonly consumed drugs in the country. This led to his famous attempt to prosecute Coca-Cola for mislabeling, in which he tried have its label say that it was a habit-forming concoction containing cocaine, which was not yet a banned substance. When he was unable to support this claim (the traces of cocaine it originally contained had been removed decades before), he attacked it for its high caffeine content. Caffeine, he said, resembled strychnine and morphine, and was therefore similar to habit-forming opium and cannabis.[48]

Attacking caffeine in a country in which coffee was practically the national drink was not a particularly wise way to rally public support. Some of Wiley's other crusades were equally puzzling. In 1908 he had tried to limit the term "ice cream" to products containing only cream. He also tried to prohibit baby food producers from calling products with any starch in them "baby food" on the (specious) grounds that children under the age of two and a half could not digest starch.[49] For years he fought a highly publicized battle to prevent whisky producers from using the term "whisky" for blends of different whiskies.[50] Now, in 1911, with his job under threat, he embarked on a campaign to prevent politically powerful corn processors from calling their sweetener corn syrup. (He wanted them to label it glucose.)[51]

The Ides of March

As evidence undermining his claims about the dangers of ingesting chemical preservatives mounted, Wiley altered his tune, emphasizing how they could camouflage "spoiled" foods that wrought havoc on the digestive system. However, critics were quick to point out that the traditional preservatives of which he approved, such as vinegar and salt, were much more effective at masking spoiled foods than benzoate of soda and other chemicals.[52]

Wiley did score a victory of sorts on one front: the Remsen Board's second decision on additives, issued in April 1911, supported his contention that saccharin should be banned, declaring that it could be "harmful to digestion" if taken over long periods of time. Wiley refrained from crowing about this, probably because the decision angered a number of powerful food companies, thereby undermining his contention that the board was biased in their favor. Moreover, the fact that chairman of the board, Ira Remsen, was the scientist who discovered how to produce saccharin out of coal tar, and that his German co-discoverer was making a fortune out of commercializing it, reinforced the board's reputation for integrity.[53]

More important to Wiley was a German government report that said benzoate of soda should not be used as a preservative because such preservatives give "the appearance of freshness to foodstuffs that already have entered upon decomposition and the buyer may be deceived as to quality." Some months later, H. J. Heinz persuaded the National Canners Association to come out against benzoate of soda, something that was easy to do since few of them now used it.[54]

Still, Wiley's repeated attacks on benzoate and the Remsen Board, as well as his refusal to let whisky and corn syrup manufacturers off the hook, finally emboldened his enemies in government to try to get rid of him. In the summer of 1911, aided by Frank Dunlap, his bureau's deputy chemist, they persuaded the attorney general to recommend Wiley's dismissal on the grounds that he had indulged in some bureaucratic hanky-panky to hire a chemist for a larger per diem payment than was authorized. Wiley's supporters in Congress then called a committee hearing at which Wiley railed against his enemies in the administration who, he charged, had suppressed his critical reports on dangerous additives and had given his bureau's power to ban them to the Remsen Board. To justify his position, he trotted out the German government report on benzoate. Faced with this strong show of support from Progressive congressmen, President William Howard Taft let him off with a mild reprimand over the chemist's salary.[55]

Heinz and the heads of some other large food companies now demanded a "reorganization" of the Department of Agriculture that would make Wiley "supreme in food matters." However, quite the reverse happened, as Wiley's enemies in the government intensified their campaign to subvert him. Finally, in March 1912, after being forced, he said, "to come into daily contact with the men who plotted my destruction," Wiley resigned. His resignation letter evoked the twin dangers of poisonous additives and adulteration. The power to ban dangerous additives such as benzoate of soda was taken from him, he said, and given to the Remsen Board. He was also forced to abandon his attempts to forbid the sale of "so-called whisky" and of glucose masquerading as corn syrup.[56]

Hitherto a lifelong Republican, Wiley resigned from the party and began campaigning vigorously for Woodrow Wilson, the Democratic candidate in the 1912 presidential election. He attacked Wilson's two opponents, former president Roosevelt and President Taft, for allowing the continued use of dangerous food additives. He recounted the story of how Roosevelt blew his top over saccharin and tried to stop him from warning of the dangers of benzoate of soda and saccharin, something Wiley called "a crime against humanity." Taft, he said, had ordered key prosecutions to be dropped.[57]

Wilson, in return, backed Wiley's campaign against benzoate of soda. Although he had been a professor at Johns Hopkins and president of Princeton, he declared that he had no confidence in boards of academic experts such as the Remsen Board, for they looked at problems too narrowly. It had concluded correctly that benzoate of soda in small quantities was not harmful, but it had failed to ask whether, as Dr. Wiley charged, it was being used to mask products that had "gone bad."[58]

Four days earlier, though, Wiley's campaign against benzoate of soda had received another major blow. A meeting of the nation's top nutritional scientists approved the report of Dr. John H. Long, a chemist at Northwestern University, which said that a Poison Squad of his students had been fed the chemical for one hundred days with "absolutely no ill effects." Wiley could only respond with a cryptic analogy: "The people ask for bread," he said, "and Dr. Long and his assistants give them a stone in the form of the moribund benzoate."[59]

By then, so many Poison Squads had been put to work that the term had lost its impact. Wiley himself had contributed to this. He had created one to test soda fountain drinks and "nerve tonics" to see if they contained opium, cocaine, or caffeine. Another tested headache powders. When baby foods came under his scrutiny, there were rumors that he was going to create a Poison Squad of babies. This was by no means far-fetched, for he did set up a "Dog Poison Squad" to test dog food. (The experiment was stopped prematurely when the bureau's neighbors complained about the nighttime howling.)[60] Also, as we have seen, Wiley's critics set up their own Poison Squads to refute him.[61] Dr. Long used a squad to test one of Wiley's targets, copper sulfate, which made canned green peas look green. He reported that not only did it not have any ill effects; it "could be considered a desirable component of French peas."[62] In the end, no member of any of the squads, not even a dog, was ever really poisoned, at least in the conventional sense of the word.

Nevertheless, for about eleven years the public had heard the highly respected official watchdog of food safety repeatedly warn that food processors were using poisons in their products. The long-term effects of this, while impossible to gauge, could not have been negligible.

Shadows of Former Selves

In the short run, though, the failure to find any "smoking guns" on additives helped dissipate fears of them. So did the federal government's abandonment of the search for them. Wiley's successor as chief of the Bureau of Chemistry, Dr. Carl Alsberg, said the bureau's role was to "help honest manufacturers to discharge their duty by supplying wholesome products. The Bureau of Chemistry belongs not only to the consumer but also to the manufacturer." This meant concentrating its enforcement efforts on small producers who mislabeled and adulterated food.[63]

Woodrow Wilson did nothing to change this. Once he assumed the presidency in March 1913, he lost all interest in benzoate of soda and "pure food." Then, in 1914, the Supreme Court effectively declawed the bureau. It ruled that the Pure Food and Drug Act's prohibition against using "any added poisonous ingredient or other added deleterious ingredient which may render [food] injurious to health" meant that the government could not ban a chemical just because it was dangerous in some circumstances. It had to prove that it injured humans when it was added to specific foods in specific quantities.[64]

Some observers predicted that the 1914 Court ruling would lead the bureau to revive its Poison Squads, but this did not happen. Instead, it drew even closer to the companies it was supposed to be supervising. In 1920, when bottled California olives infected with botulism killed a number of people across the country—including all but one of an Italian American family in the Bronx—the head of the bureau's food division rushed out to California, not to prosecute the offending processor but to reassure the public. Suspect bottles of olives were removed from grocers' shelves, and the bureau set up a special Botulism Commission in California to ensure that it would never happen again. In the years that followed, the *entente cordiale* between the bureau and the industry grew even closer, as bureau scientists left their jobs for ones in industry, creating the kind of revolving door that would later characterize the so-called "military-industrial complex."[65]

Ultimately, as James Whorton concluded in his study of pesticides, the chief problem with the Pure Food and Drug Act was not weaknesses in the act itself. Rather, it was resistance to vigorous enforcement within the Department of Agriculture, which was committed to promoting the interests on the very food producers who felt threatened by such enforcement.[66] Moreover, during the 1920s, when Big Business was hailed as leading America into a "New Era" of never-ending prosperity, there seemed to be nothing untoward about Wiley's old unit, which was renamed the Food and Drug Administration, cooperating with large food processors to ensure the safety of their products.[67]

Wiley himself followed a trajectory similar to that of his ex-department. He became the director of the Bureau of Foods of *Good Housekeeping*, which gave its "Seal of Approval" to its advertisers' products. Given the large number of food advertisements in the magazine, one might expect that he would have had some difficulty repeatedly writing, "You may safely rely on those articles of food which are advertised in *Good Housekeeping*."[68] But this seems to have posed him no problem, perhaps because most of the magazine's corporate advertisers were exactly the kind of large companies that had originally supported him. In his monthly column, he occasionally condemned adulteration, but, as in the past, he aimed at the kind of chicanery that involved small-timers: butter adulterated with water, whole wheat flour mixed with other grains, and so on. There is no record of him ever recommending that the magazine's Seal of Approval be withheld from any food advertiser. It was awarded to Jell-O and other foods of questionable nutritional value, as well as to Fleischmann's Yeast, which made outlandish claims that its yeast cakes cured such things as acne, tooth decay, "fallen stomach," and "underfed blood."[69] Occasionally, Wiley did raise the benzoate of soda issue, but other old causes remained dormant. His concerns about the deleterious effects of caffeine disappeared as rapidly as the magazine's ads for coffee proliferated. In 1928, responding to a reader's question about the use of sulfur dioxide in dried fruit, he now applauded it. Not only did it preserve the fruit's color, he said; it killed harmful insects as well. One wonders whether this reversal was connected to the many ads the magazine carried for California Sun-Maid dried fruits.[70]

With regard to diet, Wiley's columns recommended "moderation in everything," and it certainly seemed to work for him. He married for the first time in 1911, at age sixty-seven, fathered two children, and kept working at *Good Housekeeping* until shortly before his death in 1930 at age eighty-six.

By then, the country was spiraling down into the Great Depression, and public sentiment was again turning against big business. Once again, critical eyes were cast on what it was doing to the food supply. In September 1933 the Food and Drug Administration mounted an exhibit called the "Chamber of Horrors" that gave horrific examples of dangerous, sometimes lethal, substances that it had no power to ban. Although most were cosmetics, some involved foods, including questionable products of the butter industry and sprays used on fruits and vegetables. The first lady, Eleanor Roosevelt, emerged from the exhibition disgusted and alerted her husband, Franklin, to the need for changes in the law.[71] Then Arthur Kallet and Frederick Schlink's best-seller *100,000,000 Guinea Pigs: Dangers in Everyday Foods, Drugs, and Cosmetics* revived many of Wiley's charges about chemicals in food. Originally entitled *Poisons for Profit*, it warned that "poisons" such as sulfur dioxide and lead arsenic were routinely sprayed on fruits, and it condemned benzoate of soda and sulfites as dangerous chemicals that camouflaged rotten foods.[72] A mini-wave of so-called "guinea pig journalism" followed, raising the same kinds of suspicion of processed food as had the "muckrakers" of the early 1900s. As a result, in 1934 President Roosevelt's secretary of agriculture, Rexford Tugwell, sent a new food and drug bill to Congress, which would give the FDA the power to prevent the use of poisonous food additives and prohibit false health claims for foods.

What ensued represented a striking contrast with Wiley's campaign for the original pure food act. Tugwell was a combative Columbia University economics professor who, unlike Wiley, was unwilling and unable to persuade major food producers to support it. As a result, industry representatives in Congress easily fought the bill to a standstill. The Republican-dominated press attacked "Rex the Red" Tugwell, calling the bill Communist-inspired. Women's magazines such as *Ladies' Home Journal*, which had forcefully supported Wiley, responded to food advertisers' pressure and joined the red-baiting chorus. The powerful Hearst chain, which had published Wiley's column in *Good Housekeeping*, ignored the pleas of Wiley's widow and joined in. Only a particularly horrendous poisoning scandal, in which a Tennessee druggist who produced a patent medicine that killed seventy-three people was given the maximum penalty under the old law—a $200 fine for false labeling—finally spurred passage of the new Food, Drug, and Cosmetic Act in 1938. By then, it had been quite thoroughly disemboweled.

The act's weakness in regulating the food supply was exemplified soon after it passed. The first major action against foods that made false health claims—one that forced Fleischmann's to stop making ridiculous claims for its yeast—was initiated not by the Food and Drug Administration, but by the Federal Trade Commission, which regulates advertising. Meanwhile, all of the chemical additives currently in use—including Wiley's *bête noire*, benzoate of soda—were included in a list of additives that the

FDA deemed (without actually having tested them) "generally recognized as safe." For the moment, though, this was hardly noticed. Fear of what was being added to foods had now been overshadowed by the opposite concern about food processing: that it deprived food of its healthful qualities.

6

Vitamania and Its Deficiencies

Elmer McCollum's Rats

One of the (literally) most distasteful aspects of my otherwise happy 1940s childhood came with the approach of winter, for with it came the dreaded daily dose of fishy-tasting, oleaginous cod liver oil, the era's main conveyor of vitamin D. In addition, when I was judged to be too short and skinny, I was practically force-fed fistfuls of pills containing other vitamins. Little did my doting mother know that she was responding to one of the twentieth century's most persistent concerns: fear that food processing was depriving food of essential nutrients.

How did my mother come to worry about vitamins? She only had about three years' education in a shtetl school in pre–World War I Poland and two or three additional years in a primary school in Toronto. However, by the time I was growing up, she read the daily newspaper, subscribed to *Time* magazine, and spent much of her day listening to the radio. These must have been the main sources of her concerns, for by the 1930s and 1940s the media were full of warnings about the dire consequences of not having enough of these essential food components.

These fears dated back to the discovery of vitamins in the 1910s and the rise of the so-called "Newer Knowledge of Nutrition." Before then the ruling paradigm was the "New Nutrition," which had come along in the 1880s. It regarded the human body as analogous to a steam engine and saw food as composed of carbohydrates and fats, which provided fuel for the human machine, and proteins, which kept its muscles in good repair. Scientists calculated how much fuel, measured in calories, was needed to keep the system going, and how much protein, measured in grams, would prevent it from falling into disrepair. They then analyzed foods to see how much of these elements they contained and tried to match daily food intake to the "machine's" needs.

The American scientists who helped construct this system were celebrated in the nation's press. Home economists and social workers taught its lessons in schools, women's clubs, and government agencies. By 1910, then, the middle-class public was accustomed to the idea that food contained essential substances whose presence could be determined only by scientists in laboratories. They were also on board with the idea that science, not taste, was the best guide for judging the healthfulness of food. It was relatively easy, then, for them to accept the idea at the core of what I have called "the Newer Nutrition," which said that food also contained invisible elements called vitamins that could also be detected only in the laboratory.[1]

Who actually discovered vitamins was, and remains, a controversial subject. However, there can be no doubt that the idea that they were essential to life owed much to Casimir Funk, the young Polish biochemist who gave them their name. By the early 1900s, a number of chemists suspected that good health depended on much more than the New Nutrition's holy trinity of proteins, fats, and carbohydrates.[2] But it was not until 1911 that Funk, working at the Lister Institute in England, managed to isolate a water-soluble "accessory factor" whose absence in the diet caused beriberi, a disease that was common in East Asia. In what proved to be a world-historical stroke, he christened it a "vitamine."

In 1915 the American chemist Elmer McCollum isolated another substance whose absence caused an eye disease in rats and seemed to stunt their growth. At first, he called it "fat-soluble A," but, realizing that Funk had come up with a much catchier name, he changed it to "vitamine A," deftly relegating Funk's discovery of four years earlier to second place, as "vitamine B." This was followed, over the next seven years, by the isolation of vitamins C and D, which were found to prevent scurvy and rickets, respectively.[3] These discoveries received breathless attention in the press, and although these ailments were hardly present in America, the nation was soon swept up by a wave of concern regarding the healthfulness of its food that was later labeled "Vitamania."[4]

As early as 1921, when only a few vitamins had been discovered and little was really known about them, Benjamin Harrow, a prominent doctor, warned the public that an insufficiency of them resulted in diseases that "manifest symptoms which are hideous, revolting. . . . Millions have died for lack of vitamines."[5] Why did such alarums strike such a sensitive note among the public? Harrow himself gave the most important answer when he said, "The very term is pregnant with meaning."[6] Surely my mother would not have been so anxious to ply me with them had they been saddled with one of the twenty-three other names that scientists initially suggested; that is, names such as "fat-soluble A," "water-soluble B," "anti-scorbutic C," "accessory food substance," and "food hormone."[7] The genius of the term "vitamin" is that it is double-barreled, connoting something essential for both life *and* vitality. McCollum, for example, played on this by calling them essential "for the preservation of Vitality and Health."[8] Not only did my mother expect them to keep me healthy; she also hoped they would give me the energy to practice the piano.

As with germophobia, solid scientific backing helped vitamania trump the panoply of other food fads and nostrums tempting Americans. Leading the charge was the folksy chemist Elmer McCollum. Raised by a very supportive mother on an isolated Kansas farm, he progressed from a one-room rural schoolhouse to graduating showered with honors at the University of Kansas. He was then accepted for graduate work at Yale, where he did his doctorate in chemistry under the prominent nutrition researcher Thomas Osborne. Yale proved to be a bump in the road, though, for McCollum felt that high-flying nutritional scientists such as Osborne and his colleague Lafayette Mendel looked down upon him as a midwestern "hick." This seemed confirmed when, after completing his degree in 1906, McCollum failed to secure a post as a researcher at a prestigious Ivy League university. Instead, he had to return to the Midwest, to a position at the state agricultural experiment station in Madison, Wisconsin, adjacent to the state university, which did its nutrition research on cows.[9]

But McCollum's ambition remained undimmed, and he set out to make his mark by proving the existence of the suspected "accessory factors" in food. He became a fixture at scientific meetings, displaying photographs comparing scrawny, debilitated cattle with healthy ones that had been fed different feeds that contained exactly same amounts of proteins, carbohydrates, and fats. He became one of the first scientists to experiment with rats, who have a much faster life cycle than cattle, and showed that there was some component in butter fat and egg yolks that was necessary for their growth and well-being. In 1915 he isolated it and, as we have seen, cleverly called it "vitamine A."[10]

In 1917 McCollum finally made it back East, as chairman of the biochemistry department of the new school of public health at the Johns Hopkins University in Baltimore, Maryland. Although the post in a public health school was not all that prestigious among scientists (his few doctoral students were mainly women in home economics), the opportunities for self-promotion in the media were much better there than in Madison. He now began cultivating a Horatio Alger image as a humble Kansas farm boy who had risen to become the nation's leading nutritional scientist. In 1918 he came out with a short book explaining vitamins to laypeople called *The Newer Knowledge of Nutrition*. In it, he slyly implied that it was he who had discovered them, something he embellished in subsequent editions. He gained particular satisfaction from telling how the Yale researchers Osborne and Mendel had furiously pursued the same goal, but had ended up down blind alleys.[11] By the time his lab isolated vitamin D in 1922, the American press was already describing him as "the famous biochemist." He went on to bill himself as the discoverer of "two of the four known vitamins." By 1925 not only was he unabashedly calling himself the discoverer of vitamins; he was also claiming, unfairly, that Osborne and Mendel had resisted his discoveries.[12]

Much of McCollum's fame derived from his warnings that insufficient vitamins in the human diet led to horrific health consequences. These he would illustrate with gruesome photographs of rats afflicted by vitamin deficiencies—skinny wrecks with coats of patchy hair who were weak, blind, and barely mobile. He would invite journalists to his lab, which could accommodate 3,000 rats, and show them two groups of rats, one of which had been fed a vitamin-deprived diet and were obviously sickly. A typical report from one of these visits said that the rats on the restricted diet would "begin to fail. Nervousness and irritability, bodily emaciation, then death would be their portion, while the other rats would remain plump, strong, bright-eyed and serene until old age comes upon them." These would often be accompanied by photographs of the healthy and the scrawny, debilitated rodents. Thanks to such stories, vitamins soon became associated with death-defying qualities. *Collier's* magazine headlined a 1922 article on them: "How Long Do You Want to Live?" A chapter in a book explaining vitamins was called "Increase the Length of Your Life."[13]

It did not take long for food producers to realize that they now had an invaluable new marketing tool: they could claim their products contained invisible, tasteless, weightless substances that were essential to the most important functions of life. Better still, although scientists could tell which foods contained vitamins, it would be many years before they could measure their vitamin content accurately or estimate how much of each vitamin was necessary to prevent deficiencies. As a result,

Elmer McCollum, the celebrated vitamin researcher, posing in 1926 with some of the rats he used to demonstrate the horrible effects of vitamin deprivation. Alarming photos of scrawny, vitamin-deprived rats on death's door played a major role in promoting "vitamania." (The Johns Hopkins University)

practically anyone, even producers of chocolate bars and chewing gum, could advertise them as packed with vitamins.[14]

Profound social changes also provided the food producers with a ready audience among women. For middle-class mothers, the 1920s produced an almost perfect storm of anxiety-producing factors. The rise of "companionate" marriage increased the pressure on wives to be their husbands' best friends and lovers. Yet at the same time, the culture was growing more child-centered, demanding that they be perfect mothers as well. Meanwhile, the servants who had been common in their households before the war practically disappeared, leaving them entirely responsible for meeting these sky-high expectations. To make matters worse, the Newer Nutrition made the traditional source of advice on feeding children, their own mothers' ideas, seem out-of-date, if not downright dangerous. These insecurities were heightened when their children went to school and returned home with warnings about the importance of those mysterious vitamins.[15]

The bane of my childhood, cod liver oil producers, were early adopters of the practice of scaring mothers. At first, they exploited Elmer McCollum's recommendation to drink cod liver oil for its vitamin A, a deficiency of which he said caused eye disease and rickets. Then, in 1922, he discovered that rickets was actually caused by a deficiency in vitamin D, which fortuitously was also present in cod liver oil. Rickets was by then quite rare in America, so he came out with a more general warning that vitamin D was necessary for early bone development.[16] This prompted cod liver oil vendors to shift into high gear, telling mothers that "scientists [say] children with an insufficient amount of the 'sunlight' vitamin . . . do not develop healthy bones and sound teeth." The Squibb company warned that, "Doctors say a startling percentage of babies' X-ray pictures show a failure of bones to grow perfectly," a problem that could be solved by its cod liver oil. Another of its advertisements asked mothers, "Can you keep *your* child off the casualty list? Will

Ads such as this one from 1936, saying that a child needed the vitamins in cod liver oil to avoid becoming a "casualty" of winter, created fears among mothers that their children's normal diets were not enough to keep them healthy.

you help him build strong bones, sound teeth, and a sturdy body this winter?"[17]

Scientists then discovered new dangers in an insufficiency of the vitamin A in cod liver oil. Some said a deficiency of it was the prime cause of infant pneumonia. Others linked it to the common cold and measles. Still others called it an "anti-infective agent," saying that this deficiency made children susceptible to serious infections of all kinds. The Parke-Davis drug company exploited the ensuing fears adeptly, telling mothers that the "resistance-building" vitamin A in its cod liver oil "increased their fighting power" against whooping cough, measles, chicken pox, mumps, and scarlet fever.[18]

Meanwhile, the realization that vitamins were not just confined to specific foods such as milk and cod liver oil, but were also present, in varying degrees, in most foods, was proving to be a boon for food producers. A wave of mergers and takeovers in the food industries during the 1920s produced a slew of companies with hefty budgets to promote their brand-named products. They helped spread vitamania far beyond the anxious-mothers market, with its focus on infants and children. Now, insufficient vitamins were said to threaten the health of everyone in the family.[19]

One of the first out of this gate was Fleischmann's Yeast. In the early 1920s, with yeast sales falling because of the steep decline in home-baked bread, it began marketing its yeast cakes as "the richest known source of" vitamin B. It said eating three of the cakes each day, either spread on crackers or dissolved in drinks, had "a truly remarkable effect on the digestive system."[20] Over the next fifteen years, the company came out with such an impressive number of dubious claims for these unappetizing cakes—from curing acne and boils to combating "nervous exhaustion" and the common cold—that, as we have seen, in 1938 the Federal Trade Commission finally charged it with false advertising.[21]

Vitamin C, isolated in 1921, also had legs, even though for years its only known property was to prevent scurvy, which was practically nonexistent in America. The California Fruit Growers Exchange, using the Sunkist brand, mounted a highly successful national campaign extolling the "daily orange" for its "health-giving *vitamines* and *rare salts and acids*."[22] By 1932 it was citing scientists' advice to drink two full eight-ounce glasses of orange juice a day, with the juice of half a lemon added to each glass.[23] (When combined with the simultaneous calls for drinking a quart of milk day and the popular penchant for coffee drinking, this must have put considerable pressure on the national bladder.)

Ironically (or hypocritically), since they themselves were food processors, some companies fed fears that modern food processing robbed foods of many of their essential nutrients. In 1920 Fleischmann's urged eating its yeast cakes because "the process of manufacture or preparation" removed from many foods the "life-giving *vitamine*" that provided the energy people needed. "Primitive man," it claimed, "secured an abundance of vitamines from his raw, uncooked foods and green, leafy vegetables. But the modern diet—constantly refined and modified—is too often badly deficient in vital elements." One of its 1922 ads said, "A great nutrition expert says we are in danger because we eat so many artificial foods—use so many of the things which are convenient under modern conditions but have been robbed of valuable properties in manufacturing." A cod liver oil producer warned that the winter months were called "the danger months" because of the scarcity of vitamins A and D in the diet. Scientists said this was because canned fruits and vegetables, "the common food sources of these precious vitamins, often lack the health giving elements they so abundantly furnish in the summer."[24]

McCollum, meanwhile, had devised the clever phrase "protective foods" to describe vitamin-rich foods. Foremost among these were milk and leafy green vegetables. According to him, milk made "all the pastoral peoples of the past and present," from the biblical Israelites to present-day Bedouins, "superior in physical perfection" to all other people.[25] He recommended that Americans boost their milk consumption from half a pint per day per person to one full quart.[26] As for leafy green vegetables, not only would these provide vitamins; they would also "serve to maintain the intestinal tract through promoting elimination."[27] He therefore recommended that Americans eat two salads a day. This was sweet music to large California corporate farmers, who were soon shipping long trainloads of their recently developed iceberg lettuce throughout the nation. By 1930 McCollum was being credited with having transformed the American diet. That year the *New York Times* recalled the impact of his photographs of vitamin-deprived rats. It said that without the vitamins provided by milk and leafy green vegetables, "the rats were stunted and nervous; they aged and died early. With these elements they grew large, muscular, and glossy coated, and lived to be rodent Methuselahs." As a result, it reported, milk sales had soared and "lettuce, formally a vegetable Cinderella, within a decade occupied the center of the grocer's stalls, with exactly seven times its previous popularity."[28]

"Wholesome" White Bread

In 1923 the women's monthly *McCall's*, clearly hoping to outdo *Good Housekeeping* with a more cutting-edge columnist than Harvey Wiley, hired McCollum to write a column on diet and nutrition. Had this been his only venture into the commercial world, one would be loath to ask whether financial considerations might have played a role in the advice he dispensed. After all, unlike Wiley, he did not have to endorse all the food products advertised in the magazine. Moreover, few people then thought there should be firm line separating the experts who told Americans which foods to eat and the companies that produced the foods. The nation's home economists—who controlled nutrition education in the schools, colleges, and government—depended on food producers to subsidize their journals, conventions, and teaching materials, as well as to employ a large number of them. No one except the margarine producers complained about the eminent biochemistry department at the University of Wisconsin being completely beholden to the dairy industry and working diligently to prevent margarine producers from using its discovery of how to irradiate food with vitamin D.[29] As a result, no eyebrows were raised when McCollum climbed into bed with a number of large food producers.

McCollum's initial steps in this direction derived from a mind-set that was similar to Wiley's. Both men were critical of the small food producers who exploited legitimate food fears to sell dubious products. In McCollum's case, it was fly-by-night producers of "vitamin concentrates" with little or no vitamin content who provoked his wrath. He urged people to get their vitamins from food, rather than from pills and potions. "The place to get vitamins," he said, "is in the market, in the grocery store, from the milk man and from the garden, not from the drug store."[30]

Needless to say, this message endeared him to a number of large food producers. First and foremost among them were the giant flour millers. During the 1920s, the trend toward eating less bread had them staggering under the weight of unsold surpluses and falling prices. This was compounded by the popularity of reducing diets, as slim figures became the ideal among middle-class men and women. Initially, the millers turned to mass advertising to pull them out of the mire. But their campaign, which featured billboards calling on Americans to "Eat More Wheat," was hardly compelling. Then they became more sophisticated. The General Mills corporation invented "Betty Crocker," a hugely successful fictional character who responded "personally" to housewives' questions about using its flour. In the late 1920s, it began using both celebrity and scientific endorsements to promote white bread. Beautiful female models and Hollywood stars were paid to extol white bread's effectiveness in weight-loss diets while male scientists and doctors testified to how it was absolutely necessary in a healthy diet. Leading the latter was Elmer McCollum.

This represented an extraordinary turnabout on McCollum's part. Since the time of the Reverend Sylvester Graham, people had been warning that milling removed healthful elements, particularly bran, from white flour.[31] Newer Nutritionists such as McCollum himself had expanded the indictment, charging that a new roller-milling technique introduced in the 1880s also deprived it of vitamins. In 1920 he wrote that not only were the proteins in white flour inferior to those in whole wheat flour, but that white flour was also "very poor" in vitamin content. Millers favored it, he said, because it could be stored longer than healthier whole wheat flour and could be shipped over long distances. Housewives liked it because they erroneously associated its whiteness with purity. "There is no justification for the demand for white flour by the public," he said. "It has been created artificially for commercial reasons."[32] Two years later he called the roller-milling process "the most important because the most insidious" of the past century's changes in the American diet.[33] It was "remarkable," he wrote, that "the most important energy food in the American diet, white flour, is notably deficient in more dietary factors than any other single food entering the diet except sugars, starches and fats which are marketed in the pure state."[34]

In September 1923, though, the National Bakers Association invited McCollum to their convention to advise them on what to do about white bread's plummeting reputation. His initial suggestion was that they add skim milk powder to their bread to make up for some of its deficiencies, something that some of them began to do. But he soon began singing a different tune. While acknowledging that whole wheat flour was "superior in its dietary qualities" to white flour, he said it was not suitable for the modern, highly centralized milling industry because it spoiled quickly. People should therefore eat white bread, which was "a good source of starch (an energy food) and proteins," and supplement it with milk and green vegetables, the "protective foods."[35]

Surely not coincidentally, General Mills, the largest of the millers, then hired McCollum as a spokesman, and he became even more positive about white flour. He now began warning that whole wheat bread eaten alone was not at all superior to white bread consumed with milk or leafy vegetables.[36]

Before long, other medical experts were following McCollum's lead by denigrating whole wheat bread as less digestible and therefore less nutritious than white bread.[37] The American Medical Association came aboard, officially endorsing white bread as "a wholesome, nutritious food with a rightful place in the normal diet of the normal individual." In 1930 the U.S. Department of Agriculture and the Public Health Service lent their backing to the campaign by calling both white and whole wheat bread equivalent "economical sources of energy and protein in any diet."[38]

McCollum's carefully crafted reputation as the man who discovered vitamins suffered a setback in 1929, when the committee that awarded the Nobel Prize in Medicine for the discovery of vitamins ignored him completely.[39] Nevertheless, Americans' belief that he was the world's leading vitamin expert remained unshaken, and his value as a corporate spokesman was undiminished. General Mills' advertisements continued to feature him praising white flour's "wholesomeness." He extolled it in a "Betty Crocker" pamphlet called *Vitality Demands Energy—109 Smart New Ways to Serve BREAD, Our Outstanding Energy Food.* In 1934 he joined a bevy of Hollywood stars (as well as Lafayette Mendel, who still despised him for claiming to have discovered vitamins) in a special radio program hosted by "Betty Crocker." With a full orchestra playing in the background, he spoke of "the scientific arguments in favor of eating bread." To top it off, the next year he wrote a public letter to a congressional committee to condemn "the pernicious teachings of food faddists who have sought to make people afraid of white-flour bread."[40]

The Great Acidosis Scare

Years later, in his autobiography, McCollum justified his defense of white bread by pointing out that he had suggested adding skim milk powder to it and said that it should be consumed along with "protective foods." However, he did not even mention, let alone justify, his using his scientific credibility to help other large food producers foment fear of a practically non-existent disease: acidosis.[41] A brief look at his role in this helps explain why this gap was no inadvertent oversight.

It is not that there was no such thing as acidosis—it is a rare blood ailment that can severely debilitate diabetics. Before the mid-1920s, some doctors claimed to have seen it in non-diabetics, but this was mainly in infants whose diets contained too much fat. However, as dieting to lose weight swept the middle class, some physicians began diagnosing acidosis in adults. They opined that extreme reducing diets forced the body to burn too much fat, causing an imbalance of acid in the bloodstream and the kind of acidosis experienced by diabetics. Other experts, citing widespread complaints of acid stomachs, then expanded the indictment. Excess acid in the stomach, they said, had made acidosis rampant among a host of people who were not on extreme diets, with dire consequences for their health.[42]

Leading this new charge was Elmer McCollum, who joined the battle with a column in *McCall's* in February 1928. Until then, the main symptoms of acidosis were said to be fatigue and lassitude. Now, according to McCollum, they included

> lassitude, malaise, nausea, sometimes vomiting, headache, sleeplessness, weakness, and loss of appetite. The muscles ache, the mouth becomes acid with resulting injury to the enamel of the teeth.... Some eminent physicians now believe that the disease of the blood-vessels which is responsible for high blood pressure, kidney diseases, gangrene, and apoplexy are the result of prolonged injury due to eating excessive amounts of acid-forming foods.

As a coda, he also attributed "much of the chronic fatigue" that seemed to affect businessmen and housewives to "diets that are too acid."[43]

Others added acidosis in children to the list of mothers' worries. The *Washington Post*'s medical adviser said, "The child with acidosis is dopey, drowsy, and acts sick. He complains of headaches and the thermometer shows he has fever. His breath is foul; this breath odor has been compared to the sweetish smell of overripe apples. The urine is highly acid. When tested, it is found to contain acetone." *Good Housekeeping* told mothers to suspect acidosis when a child suffering from fever, vomiting, or diarrhea "becomes relaxed, too quiet, limp or drowsy."[44]

One would think that pinpointing too much acid in the stomach as the villain would mean big trouble for the large California and Florida citrus fruit producers. But here again McCollum was ready to lend a hand. Using the nutritionists' mantra, he advised listening to scientists, rather than one's palate, regarding which foods caused an excess of acid in the blood. When the acids in citrus fruits arrived in the stomach, he said, they quickly turned alkaline, the opposite of acid. "Foods which taste acid do not necessarily form acids in the body. . . . Even lemons and grapefruit, which taste so strongly acid, are actually alkalizers in the body." Indeed, he said, citrus fruits were among the most effective of the acid-fighters. What, then, caused the surplus of acid in the stomach? It was non-acid foods such as beef, pork, bread, and eggs that were the acid producers.[45]

McCollum's warnings were reported in major newspapers and magazines, helping to ratchet up the scare. A piece in the *New York Times Magazine* said that acidosis had now displaced vitamins as the major nutritional concern. "Where once we prated about calories and vitamins," it said, "we are now concerned with an alkaline balance." The nutrition columnist for the *American Journal of Nursing* wrote that constipation and acidosis "caused more ill health" in America than any other condition, and repeated McCollum's advice to combat it with plenty of citrus fruits and lettuce.[46]

Of course, citrus fruit marketers jumped at the chance to promote McCollum's ideas. Ads for Florida grapefruit producers said that grapefruit was what "the body needs to combat the common foe, Acidosis, for citrus fruit turns alkaline in the human system." In an advertisement for Sunkist lemons, the California growers' organization repeated McCollum's warning that acidosis was caused by "good and necessary foods such as cereals, bread, fish, eggs and meat—all of which are of the *acid-forming* type." Luckily, it said, there was "a seeming paradox": this was that although called "acid fruits," citrus fruits caused an alkaline reaction in the body that counteracted acidosis. The accompanying cartoon showed a manager lamenting that one of his underlings failed to "get ahead" because he lacked "punch." His problem: "Doctors call it acidosis. Oranges and lemons would do him a world of good—as they have for me."[47]

McCollum buttressed the citrus fruit producers' campaign with increasingly dire warnings about the practically fictitious ailment. He devoted a whole chapter to it in the 1933 edition of his popular book on diet and nutrition, saying it was "responsible for a great deal of ill health." He now added tooth decay to the list and recommended drinking plenty of orange juice as a cure. In 1934 *Parents* magazine said anxious mothers now described children's upset stomachs as "a touch of acidosis." "Everyone looks worried," it said. "They've heard it may have a fatal termination." The writer Kenneth Roberts, tongue firmly in cheek, said that leafing through a pile of new diet books left him "scared." "If I had acidosis," he said, "I was liable to pop off at a moment's notice with any one of a score of painful and unpleasant diseases."[48]

But the acidosis scare soon began fading. From the outset, some scientists had questioned McCollum's idea that diet caused excess acid in the bloodstream. Some simple experiments in the early 1930s backed them up.[49] A review of McCollum's book in a nursing journal in 1934 regretted "the introduction of controversial material [about acidosis] in a book intended for the lay reader."[50] Other medical experts now called the acidosis scare a fad, saying that it was a rare ailment that would not, in any case, be affected by orange juice.[51]

The citrus fruit producers, having gotten more mileage out of acidosis than they could ever have hoped for, dropped it and returned to extolling vitamin C.[52] But McCollum's warnings of the dire consequences of having an imbalance of acid- and alkaline-producing foods in the stomach lived on, particularly in variations on the old idea that it was dangerous to mix different kinds of foods in the stomach.[53] For instance, one of the diet books Roberts examined warned that if baked beans, a starch, were eaten with tomato catsup, an acid, "the resulting fermentation causes deadly acidosis." Another warned against eating corned beef hash because the beef is protein and the potatoes are starch. It was, said Roberts, "somewhat similar, in its effects on the stomach, to a lighted pinwheel."[54] As a result of such warnings, the automobile tycoon Henry Ford, a fervid believer in crackpot nostrums, ate only fruits for breakfast, starches at lunch, and proteins for dinner.[55]

A host of radio hucksters turned McCollum's concept into a gold mine of quack dietary advice. One of the airwaves' medical messiahs said that "acidosis and toxicosis are the two basic causes of all disease." Acidosis, he said, was brought on by eating bad combinations of food, especially starches and proteins at the same time. For $50, listeners could enroll in his correspondence course to learn how to eat foods separately.[56] Another radio guru, William Hay—owner of a sanitarium in the Poconos called Pocono Hay-Ven—had a similar message. His best-selling book, *Health via Food*, said that the root of all disease was acidosis caused by eating protein- and carbohydrate-rich foods together. He also threw in the promise that his acidosis-preventing diet would lead to weight loss, since acidosis sapped vitality and encouraged the accumulation of fat. Later in the century, this would reappear as the popular Beverly Hills Diet.[57]

Of course, all of this took place during the Great Depression, when government nutritionists were trying, with limited success, to use the lessons of the Newer Nutrition to help the millions of people who could not afford a healthy diet. But the middle classes were hardly affected by this kind of insecurity about food. Instead, they continued to navigate the shifting currents of faddish weight-loss diets.[58] If they did have other worries about food, it was the nagging suspicion, stimulated by vitamania, that modern food processing was robbing their food of its vitamins. As the 1930s drew to a close and what threatened to be a contagious war broke out in Europe, eminent experts, supported by the full weight of the government, helped heighten these concerns by warning that processing-induced vitamin deprivation was putting the very survival of the nation in peril.

7
"Hidden Hunger" Stalks the Land

Russell Wilder and the "Morale Vitamin"

At their annual meeting in November 1938, after roundly applauding a banker who condemned the New Deal for "taking from the few and giving to the many," the Grocery Manufacturers Association (GMA) presented Elmer McCollum with a special award. Officially, it was to recognize him as "one of the greatest leaders in the war against the diseases caused by faulty diet." Unofficially, it was a reward for defending them against those who condemned them for selling unhealthy, nutritionally deprived products. He had steadfastly argued that large food processors realized that "quality of products is the best appeal" and praised them for funding important scientific research, such as that showing (incorrectly, as it turned out) that the vitamin D now being added to milk prevented tooth decay.[1]

McCollum then welcomed a GMA pledge of a generous $250,000 a year for a new Nutrition Foundation that would, among other things, fund scientific research showing that processing did not deprive foods of their vitamins.[2] When this venture faltered, fifteen of the nation's largest food processors came to his rescue, putting up $1 million over five years to support research that would "seek new and better ways of processing and preserving food to get the maximum nutritive value from it."[3]

The food companies had good reason to fund such research, for new techniques for measuring the vitamin content of foods were demonstrating that processing was indeed robbing them of nutrients. To make matters worse, the chief beneficiaries of these revelations were manufacturers of vitamin pills and potions. Not only were they promising to supply the lost nutrients; they were now able to do so with products carrying exact measurements of their vitamin content. As a result, sales of vitamin supplements surged, even though there was still nothing like a consensus about how much of them the body needed. By 1938 Americans were spending $100 million a year on them, and producing them had become a very big, and very profitable, business indeed.[4]

Food producers countered by trotting out McCollum and other nutritional scientists to tell Americans to get their vitamins from "protective foods" like milk and canned spinach, not pills. The American Medical Association, fearing that people would turn to vitamin pills, rather than doctors, to cure what ailed them, rallied to their side. So did pharmacists, who feared that vitamins would be sold in every corner store. They all pressed the government to have vitamin supplements sold only by prescription. But the Food and Drug Administration resisted, declaring that as long as vitamins were not prescribed for illnesses or accompanied by any health claims, they were a kind of food and could therefore be sold virtually anywhere. Soon department stores such as Macy's and supermarkets such as Kroger were selling their own brands of vitamins.[5]

The outbreak of war in Europe in September 1939 seemed to sharply raise the stakes in this conflict over nutrition. By the late summer of 1940, the Nazis dominated most of the Europe and were preparing to invade Great Britain. For the United States, the possibility of being drawn into the war now loomed large. A massive rearmament program was mounted and a military draft begun. Questions now arose as to whether the Depression-battered workforce would be vigorous enough to meet the demands of industrial or military service. Government nutrition experts solemnly warned that three-quarters of the nation suffered from "hidden hunger," meaning vitamin deficiencies that were camouflaged by full stomachs and healthy appearances.[6]

Since "hidden" signified having no apparent symptoms, the nutritionists could avoid showing how the deficiencies were linked to serious diseases or death. Instead, they warned of such things as the dangers inherent in sending soldiers deficient in vitamin A, which was thought to protect against night blindness, on patrol at night.[7] The results of the physical examinations given to men in the first draft call-up in October 1940 stoked these worries. Almost 50 percent of them were rejected, mainly for physical disabilities that were said to be related to vitamin deficiencies. Only those who read far below the headlines learned that the largest number were rejected for tooth decay, which, as we have seen, was thought to be related to a deficiency of vitamin D.[8]

The draft board news seemed to confirm suspicions of the dire effects of modern food processing. Medical and public health officials now warned that the nation was suffering from a "vitamin famine" because so many foods were deprived of their vitamin content before they reached Americans' homes.[9] *Hygeia*, the American Medical Association magazine for laypeople, warned that "the trend away from the use of fresh or 'whole' foods and the substitution of preserved and refined food" was threatening the health of Americans.[10] The *New York Times* said, "The discovery that tables may groan with food and that we nevertheless face a kind of starvation has driven home the fact that we have applied science and technology none too wisely in the preparation of food."[11]

In May 1941 President Franklin Roosevelt, concerned about worsening reports of draftee rejections, called five hundred of the nation's leading food and nutrition experts to a conference in Washington to "create a framework for a national nutritional policy." During World War I, the government's food conservation program had helped propagate the basics of the New Nutrition. The next generation of nutritional scientists now sought a program that would educate the public about the Newer Nutrition, with its emphasis on vitamins. In his keynote address, Vice President Henry Wallace said people should be taught to consume more "protective foods," which would provide the nutrients to "furnish the nervous energy to drive us though to victory." Dr. Russell Wilder, chairman of the National Research Council's Committee on Food and Nutrition, warned that 75 percent of Americans were suffering from "hidden hunger, a hunger more dangerous than hollow hunger, because the sufferer, lacking in essential food elements, although his stomach might be full, was existing on the borderline between health and disease." Other speakers echoed these warnings that Americans must add additional vitamins to their diets to combat "hidden hunger."[12]

The problem was that estimates of how much of these nutrients were necessary were still all over the map. A committee of nutritionists was asked to prepare a list of "standards" for eight important nutrients—that is, the minimum amounts necessary to prevent illnesses. Instead, they produced what they called "recommended daily allowances" (RDAs). These were the average of various scientists' estimates, but with a 30 percent "margin of safety" added on. This seemed to satisfy those whose estimates were high while mollifying the low-ballers with the argument that an oversupply of vitamins could surely do no harm.[13] But the evasive term "allowances" was soon ignored, and the RDAs were almost immediately taken as the minima necessary for good health.[14] The Surgeon General presented them as a new "yardstick" to the public—"a nutritional gold standard for the American people."[15]

By then, a major villain in fostering "hidden hunger" had hove into sight: the one that McCollum had performed his flip-flop over in 1923. Once again, modern milling was accused of removing essential vitamins from flour. Leading this charge was Russell Wilder, head of the NRC's food and nutrition committee, who deftly used the new RDAs to rally the nation behind his pet project. This was to ensure that the nation had the "nervous energy" necessary to win the war by stoking the national diet with vitamin B_1 (thiamine), the "morale vitamin."

A handsome white-haired doctor with the avuncular looks associated with medical wisdom, Wilder had already convinced many of the nation's leaders by the time of the 1941 National Nutrition conference that America's very survival was being threatened by a lack of thiamine. How did he accomplish this? It all began when he and some of his colleagues at the Mayo Clinic in Rochester, Minnesota, became convinced that the nutritional quality of the American diet had deteriorated markedly over the past sixty years. The main cause, they thought, was that the new steel roller-milling process introduced in the 1880s removed most of the vitamins in the B complex, including 90 percent of the thiamine, from the white bread that constituted about 30 percent of most Americans' caloric intake.[16]

To demonstrate the baneful consequences of this, they put four young female patients at the nearby state mental hospital on a diet that was very low in thiamine. After five weeks, the researchers observed a number of symptoms, led by "anorexia, fatigue, loss of weight . . . constipation, and inconstant tenderness of the muscles of the calves."[17] This was followed by a study of six more female inmates who worked on the Mayo Clinic cleaning staff. After six weeks on the same diet, they suffered from "debility . . . fatigue and lassitude." Among the symptoms recorded were "depressed mental states, generalized weakness, dizziness, backache . . . anorexia, nausea, loss of weight," and vomiting. They also suffered from "psychotic trends" and mental problems that the researchers said closely resembled those of neurasthenia, the one-size-fits-all diagnosis of "nervous exhaustion" popular in Victorian times. When, in the midst of the experiment, thiamine was restored to the diets of two of the women, they experienced "a feeling of unusual well-being associated with unusual stamina and enterprise." All the women recovered well when they went back on their normal diets.[18]

The implications for the nation's defense effort seemed obvious: thiamine was essential for the psychological health of the nation. The publication of the study in October 1940 prompted an editorial in the *Journal of the American Medical Association* warning that, in case of invasion, the "moodiness, sluggishness, indifference, fear, and mental and physical fatigue" induced by lack of thiamine would mean the difference between resistance and defeat. Various government officials joined in, saying that the nation had to be educated about the importance of this "potent morale builder." Thiamine quickly became "the morale vitamin." No one said that this was based on a study of the mental states of ten patients in what was then called an insane asylum.[19]

Wilder's solution to the problem was to restore back into white flour the thiamine that roller milling removed.[20] This went down very well indeed with vitamin manufacturers. New techniques for measuring vitamin content were now allowing them to sell vitamins concentrates to manufacturers of a number of processed foods, including chewing gum.[21] By late 1940 General Mills—makers of Wheaties, the "Breakfast of Champions"—was already on the thiamine wagon. It advertised that it added thiamine to the product because "in the necessary process of transforming whole wheat into a delicious breakfast cereal, the vitamins and sometimes the minerals are too often impaired or destroyed." A shield on its box proclaimed that it was "Accepted by the American Medical Assn. Council on Foods," which verified that its "Nutr-a-sured" process restored the "vital vitamin B_1" and other nutrients to the wheat.[22]

The flour millers, on the other hand, balked at paying the high prices that vitamin manufacturers were demanding. McCollum tried to help them out by advising Americans to eat other foods to make up for what was lacking in white bread. He also proposed fortifying bread not with vitamin concentrates, but with inexpensive skim milk powder and brewers' yeast. The American Meat Institute had a different suggestion, advertising that one pork chop provided an individual's entire daily requirement of thiamine.[23] Perhaps the most bizarre alternative was that put forward by an Andrew Viscardi. He patented a process for giving smokers the "therapeutic benefit" of vitamin B_1 by impregnating tobacco with it.[24]

However, nothing but adding thiamine to flour would satisfy Wilder. After he gained federal government backing for this, the millers were forced to respond. In February 1941 most of them agreed to begin producing "enriched" flour for bread. (They rejected calling the flour "reinforced," because it implied, said one report, "that there was something wrong with the old fashioned bread and it needed jacking up.") The new flour included not only thiamine, but also iron, riboflavin, and pellagra-fighting nicotinic acid (later called niacin).[25]

The step was hailed as "designed to rescue some 45,000,000 Americans from hungerless vitamin famine." The millers pointed out that this was being done "the American way," voluntarily.[26] But the millers did not volunteer enough, and three-quarters of the nation's flour—much of which was not used for bread—still remained un-enriched. The media now called for making enrichment mandatory, emphasizing the importance of boosting intake of the "morale vitamin." The *Washington Post* wrote of "how vital are certain vitamins in maintaining . . . the morale of the nation." The Surgeon General said that the national defense demanded the use of thiamine to "build up the morale, pep and infection resistance of the Nation."[27]

Again Wilder took the lead. He said that the Nazis were deliberately depriving the conquered peoples of European of thiamine to reduce them "to a state of mental weakness and depression and despair which will make them easier to hold in subjection." This soon morphed into the charge that they were systematically destroying the thiamine content of all the food in the occupied nations. In April 1941 he told the College of Physicians that over the past sixty years (that is, since the new roller mills), there had been a "constantly increasing deficiency" in nutrients in the American diet, leaving two-thirds of the nation suffering from "serious malnutrition." Tests on young women at the Mayo Clinic had shown that a deficiency of thiamine in the American diet "may have led to a certain degree of irremediable deterioration in the national will."[28]

He also used the experiment to warn that thiamine deficiency was imperiling American industry. In June he told a scientists' meeting that depriving the cleaning women of thiamine had led to "drastic personality degeneration and working inefficiency."[29] In an interview for the *New York Times Magazine*, he described them as

> a number of women volunteers [who] began as sociable, contented workers on what looked like an acceptable palatable diet. In a few weeks they were quarrelsome, depressed, and easily fatigued. They even went on strikes. Forty-eight hours after thiamin was added to their deficient diet they were their old selves again.[30]

When a number of strikes broke out in defense industries, he warned that lack of thiamine in workers' diets was a major component in industrial unrest.[31]

Wilder's campaign was a resounding success. By June 1942 the millers had fallen in line, and almost all of the nation's bread was being made with enriched flour. A year later the millers complied with a government order to increase the amounts of other vitamins added to flour as well.[32]

Vitamins for "Ooomph"

A favorite tactic of thiamine's promoters was to use the vitality implicit in the name "vitamin" to claim that it combated low morale by providing "zip," "pep," and energy. At the 1941 national nutrition conference Vice President Wallace said, "What puts the sparkle in your eye, the spring in your step, the zip in your soul? It is the oomph vitamin."[33] A study by a Toronto doctor was cited as providing scientific backing for this. He reported that the ability of ten subjects on diets "rich in B_1" to hold their arms out horizontally far exceeded that of four men whose diets had not been enriched. After a week of having their diets supplemented with thiamine, the four weaklings could extend their arms as long or longer than the others.[34]

Such news piqued the interest of professional sports managers, always looking for an edge. The New York Rangers hockey team, perennial underachievers, were given large amounts of the vitamin B complex and skated their way to the Stanley Cup championship in the spring of 1941. This set the baseball wizard Branch Rickey, the general manager of the St. Louis Cardinals, to providing his players with what he called the "Wham" or "Oomph" vitamin, albeit with less success. (The Brooklyn Dodgers held them off to win the pennant that year.)[35] The food editor of the *New York Times* told housewives that the "morale vitamin" would give their families the stamina for the war effort.[36] Producers of a garden fertilizer called "Tropix B_1," guaranteed that the thiamine in it would produce giant flowers and plants. It came as no surprise that in November 1941, when the Gallup Poll asked Americans to name a vitamin they had recently heard much about, they overwhelmingly named vitamins B_1 and B_2. Significantly, 84 percent of the housewives polled could not tell the difference between vitamins and calories, indicating that they likely equated vitamins with energy.[37]

Much to the chagrin of the food producers, it was the vitamin manufacturers who reaped most of the profits from this. They persuaded employers in war industries to distribute vitamins to their workers to give them "pep" and "energy." Managers and workers reported enthusiastically about their efficacy.[38] Pratt and Whitney reported that productivity at one of its aircraft plants increased 33 percent and spoilage of materials plummeted after it began distributing small "vitaminized" chocolate cakes to its workers. A union in another plant demanded and gained a contract promising that the employer would provide workers with vitamin concentrates. Employees in less arduous industries were also thought to benefit. The entire workforce of the Columbia Broadcasting System was told to take vitamin pills daily, at company expense. By April 1942 almost one-quarter of Americans were taking vitamin pills.[39]

Wilder, meanwhile, moved on from demanding the restoration of nutrients lost in processing to seeking to have foods "fortified," which meant adding nutrients that had never been there. He proposed adding milk solids to white sugar and vitamins to lard and other fats. He even called for fortifying fresh fruits and vegetables with vitamins.[40] However, there was little enthusiasm for such schemes, particularly from food processors. Aside from bringing attention to their products' nutritional deficiencies, it would also mean considerable added expense. Nor would organized medicine support it, as the AMA continued to see vitamin supplements as a threat to their incomes.[41]

Ultimately, Wilder failed to keep thiamine deficiency front and center among the nation's wartime concerns. Despite his warnings that the deficiency persisted, the nation's morale seemed fine. New studies challenged the conclusions he drew from the Mayo experiments. They indicated that giving people added doses of thiamine seemed to have no effect at all on their morale, or anything else, for that matter.[42] By mid-1943 bucking up morale was hardly being mentioned as one of thiamine's attributes. Instead, the U.S. Public Health Service urged consuming foods with plenty of B vitamins because they provided energy and thereby prevented "involuntary laziness." A *Wall Street Journal* piece on how drug companies were profiting mightily from supplying thiamine for bread mentioned only that it prevented beriberi, which was non-existent in America. As doubts about thiamine's importance mounted, the government's nutrition advisory board finally slashed the recommended daily allowance for it.[43]

Another "Fountain of Youth" Tablet

Competition for attention from another member of the B complex also helped push thiamine out of the limelight. In 1937 and 1938, Agnes Fay Morgan, a home economist at the University of California at Berkeley, began reporting that experiments on her rats showed that it was the lack of one of the B complex's members—she wasn't sure which—that was responsible for their hair turning gray. By early 1940 she was linking a lack of this "anti-gray hair vitamin" in the diet to aging. Vitamin producers, with visions of a bonanza dancing before their eyes, quickly set their scientists to work. In September 1941 Dr. Stefan Ansbacher of the International Vitamin Corporation identified the rejuvenating substance as para-aminobenzoic acid (PABA) and presented what he said was the first evidence that it could restore color to gray hair in humans. The national press jumped on the story, reporting that PABA appeared to "aid in the prevention of aging" and presaged a "Fountain of Youth" tablet.[44] Ansbacher then released a study claiming that PABA had reduced graying in three hundred people of all ages and both sexes, and had produced "increased libido" in many of them as well. The science columnist of the International News Service promptly hailed it as one of the year's ten most important scientific discoveries, on a par with the discovery of penicillin.[45] Two large vitamin producers then tested it on eighty gray-haired male prisoners. Not only did it seem to restore the hair color of two-thirds of them; it also "markedly increased the libido in almost all cases."[46]

But attempts to market PABA soon fizzled. At first, this was because its availability was restricted—it was an ingredient in TNT and Novocain. Then when pharmacies began selling it in late 1942, customers learned that they had to take the pills four to six times a day for three or four months before expecting to see any results, and that these were by no means assured.[47] Dr. Ansbacher himself, whose hair remained steadfastly gray, was hardly a good advertisement for the product.[48]

Ansbacher was soon reduced to promoting PABA's supposed libido-stimulating qualities.[49] Other scientists took to promoting it as an anti-sunburn ointment or a cure for typhus and thyroid problems.[50] In February 1943 an experiment showing that PABA detoxified arsenic compounds was hailed by the *New York Times* as proving that its power to turn gray hair back to its natural color could no longer be doubted, but this failed to revive faith in its anti-aging properties.[51] It was rapidly submerged in a rising tide of claims for other vitamins, which scientists said cured everything from blindness to cancer. One of the more optimistic ones said that increased vitamin intake would lead to a more moral society. It argued that since "intelligence and morality" go together, and the widespread "dullness" among lower-income children was due to poor diets and lack of vitamins, increasing their vitamin intake would lead to higher intelligence and a consequent lessening of their propensity to lie, cheat, and steal. Moreover, since niacin was known to cure the "mental derangements" of pellagra victims, thereby turning people who were "crazy" back into useful citizens, it was clear that "an abundant supply of vitamins [would] promote intellectual keenness." For society as a whole, then, there was but one conclusion: "Vitamins in the future will not only give people better health, both bodily and mentally, but will increase their morality."[52]

Food manufacturers watched this kind of vitamania in despair. They talked of mounting advertising campaigns to, as one reporter said, "wrest back from the drug trade the lucrative business in vitamins." Their allies in the federal government talked of organizing a vast educational campaign to teach industrial workers how to get their vitamins through diet. But the plans came to naught, and Americans' propensity to pop pills swelled. By 1944 the vitamin producers were boasting that sales had grown "by leaps and bounds," estimating that three out of four Americans were now consuming them.[53]

Vitamania Forever?

The belief that vitamin pills would provide energy and ensure good health persisted unshaken into the postwar years, which is when my mother stuffed me with them, even though there was never any credible evidence for this. In 1946 the results of a long study comparing the condition of a group of about 250 Southern California aircraft factory workers who had been given large doses of vitamin supplements with a control group of similar size who had been given placebos appeared. To the apparent disappointment of the researchers, the differences in work habits and health outcomes were negligible.[54]

But studies such as this could not shake the entrenched idea that vitamins meant vitality. It infused the scripture of postwar health food evangelists such as Gayelord Hauser, a popular guru who made his fortune by denouncing modern processing for denutrifying food. He used Wilder's study of mental patients to support his claim that the B vitamins in wheat germ would bring vitality and head off disease.[55] In 1969 an FDA survey disclosed that 75 percent of Americans still believed that taking extra vitamins gave them more "pep" and energy.[56]

They also thought, or at least wanted to think, that they cured or prevented a wide assortment of illnesses. Indeed, in 1973, when the FDA tried to restrict vitamin sellers from making health claims for their pills, congressmen reported receiving a greater volume of outraged mail than they did about the simultaneous Watergate scandal, which forced President Nixon from office. This opposition to the FDA was rooted in the rising tide of individualism of the time, and the idea that Americans should have the right to make their own decisions regarding what was or was not good for their health. It also grew out of post-Vietnam, post-Watergate disillusionment with government. So, not only did Congress turn back the FDA's attempt to restrict health claims; in 1976 it passed special legislation that—in the words of Senator William Proxmire of Wisconsin, the liberal Democrat who led the move—would remove the government's power to "regulate the right of millions of Americans who believe they are getting a lousy diet to take vitamins and minerals. The real issue," he said, "is whether the FDA is going to play God." What a far cry this was from what his Progressive predecessors thought they were doing when they passed the original act that tried to give Harvey Wiley just those powers.[57]

In the years that followed, restrictions on vitamin sellers were steadily loosened, particularly during the Reagan presidency in the 1980s. By 2006 more than half of American adults were taking multivitamins "in the belief," said a National Institutes of Health study, "that they will feel better, have greater energy, improve health, and prevent and treat disease." It was no surprise that their use was highest among women, the elderly, those with higher incomes and more education, thinner people, and among people living on the West Coast. "The irony," said the study, "is that the poor, who are much more likely to suffer from nutritional inadequacy, are the least likely to take them." As for vitamins' ability to head off illness, it concluded that there was little evidence that vitamins and minerals, taken individually or paired, had any beneficial effect on the prevention of chronic disease.[58]

Subsequent trials of vitamins backed up the study's conclusion that the vast majority of Americans did not need to take vitamin supplements. "The News Keeps Getting Worse for Vitamins" was the headline in a *New York Times* article in late 2008. "The best efforts of the scientific community to prove the health benefits of vitamins keep coming up short," it said. Two months later it reported that although almost half of American adults used some form of dietary supplement, at a cost of $23 billion a year, "in the past few years several high-quality studies have failed to show that extra vitamins, at least in pill form, help prevent chronic disease or prolong life."[59] Of course, none of this shook vitamania at all. (Indeed, I write this after having taken my daily winter dose of vitamin D, and I must confess to occasionally wondering if all that cod liver oil my mother gave me did not indeed help me develop strong bones.)[60]

Thiamine, though, was a different story. Perhaps because it is now recognized that there is more than enough thiamine in Americans' diets, it did not even figure among the vitamins that were studied for disease prevention. However, its rise and fall as the "morale vitamin" did leave a lasting legacy: the justifications for adding it to flour, centering as they did around the detrimental effects of processing on food, reinforced fears that modern production techniques removed food's vital nutrients. This helped set the stage for the next big thing: natural foods.

8

Natural Foods in Shangri-la

J. I. Rodale and the Happy, Healthy Hunza

Worries that modern food processing robs foods of their nutritional qualities are by no means new. Russell Wilder's suspicion of modern flour milling can be traced back to the 1830s and the Reverend Sylvester Graham, the nation's first advocate of "natural foods." Graham warned that the increasing number of foods being processed outside the home contravened God's laws of health and contributed, among other things, to an epidemic of debilitating masturbation among the young. He was particularly transfixed by bread, warning that the Lord had mandated eating wheat only in its natural, unadulterated form, with its grains left whole. Baking bread with white flour, he said, "tortured it into an unnatural state."[1]

Graham's disciple John Harvey Kellogg also condemned processed foods such as white flour and refined sugar as unnatural, although he never called for shunning all of them (especially not the dry cereals he developed with his brother). However, he faded into the background during the 1920s, and so did suspicion of processed food. Instead, the large food processors were hailed as helping to bring about a prosperous "New Era" of good living and healthy diets for all. Although the Great Depression once again put big food companies on the firing line, the attacks tended to focus on their shifty practices, rather than processing itself.

Yet the idea that the modern food industries were perverting nature's way managed to persist. In the 1930s it reemerged in the work of two British colonial officials in India who created the basis for modern beliefs in the healthfulness of organic and natural foods. One was the botanist Sir Albert Howard, whose studies of Indian agriculture led him to conclude that soil that was fertilized naturally—that is, with manure or composted organic material—produced more nutritious crops than those fertilized by chemicals. These natural fertilizers, he said, encouraged the activity of organisms that created nutrients in the soil, leading another Englishman to label the process "organic farming."

The other was the man who first used the term "natural foods" in its modern form, Sir Robert McCarrison, a leading physician in the Indian Medical Service. In 1922 he told a group of American medical experts that the stomach ailments such as dyspepsia, ulcers, appendicitis, and cancer that were so prevalent among Westerners were practically unknown among many of the "uncivilized" people he had treated, who were "unusually fertile and long lived." Indeed, in one case, he said, "the ruling chief . . . took me to task for what he considered my ridiculous eagerness to prolong the lives of the ancients of his people" and suggested instead "some form of lethal chamber" to rid his community of the surfeit of old people who could no longer contribute to it. The reason for this extraordinary longevity and fertility, he said, was that their diets consisted of unprocessed "natural foods"—that is, the "unsophisticated foods of Nature: milk, eggs, grains, fruits and vegetables." These were, he said, the "protective foods" recommended by Elmer McCollum.[2]

McCarrison then did experiments with rats to show that it was poor food choices, rather than lack of food or the tropical environment, that were responsible for the deplorable health of the Indian poor. In 1936 he brought these lessons home to England, where an animated debate was taking place over the nutritional status of the British working class. In a series of public lectures, he deftly used his pictures of stunted rats to condemn the typical English working-class diet of white bread, jam, canned meat, potatoes, and sweets as just as unhealthy as the white rice-based diet of the poor people of southern India.[3] His most controversial assertion, however, was that diets composed of "natural foods" that were untouched by modern farming techniques and food processing could also prevent illness.[4]

McCarrison's problem was that although he could back up his contention about the unhealthiness of the poor British and Indian diets with his experiments with rats, he had no scientific support for his advocacy of natural foods. So, when his lectures were published, he included an article he had written in 1921 about having traveled to the remote Hunza Valley, at the foot of the Himalayas in the north of what is now Pakistani Kashmir. There, he said, he had discovered "a race, unsurpassed in perfection of physique and in freedom from diseases in general, whose sole food consist to this day of grains, vegetables, and fruits, with a certain amount of milk and butter, and meat only on feast days. . . . I don't suppose that one in a thousand of them has ever seen a tinned salmon, a chocolate or a patent infant food." Their life span was "extraordinarily long," and most of the illnesses he came across in seven years of treating them were not serious—things such as accidental lesions and cataracts. Many of the afflictions common in Europe—such as appendicitis, colitis, and cancer—were non-existent. "It becomes obvious," he concluded, that a diet of "the unsophisticated foodstuffs of nature is compatible with long life, continued vigor, and perfect physique." "Civilized man," on the other hand, "by desiccation, by heating, by freezing and thawing, by oxidation and decomposition, by milling and polishing," had eliminated the natural from his food and substituted the artificial, bringing on a host of health problems.[5]

The description of the Hunza struck a particular chord in the United States. Many people there were already entranced by James Hilton's best-selling book *Lost Horizon*, published in 1933, and the popular 1937 Hollywood movie based on it. These told of Shangri-la, a utopia in a Himalayan valley, closed off from the rest of the world, where people lived to be hundreds of years old in perfect health and harmony. Word soon spread that Hilton's tale had been inspired by a traveler returning from the Hunza Valley.

However, another impetus for interest in the Hunza came from a more unlikely source—an ex-Internal Revenue Service accountant in New York City named Jerome Irving Cohen. Born in 1898 in the Jewish ghetto in the city's Lower East Side, the short, sickly Cohen, who spent his spare time tinkering with inventions, seemed destined for a life as an obscure nerd. However, in the late 1930s, with the country still mired in the Great Depression, he quit his secure government job and reinvented himself. He changed his name to Rodale and set up a company to manufacture a new heating pad that he had invented. The device had a few bugs, though, and the company struggled. (Later he would tell of a customer who had complained that one of them had electrocuted his wife but said, "If you give me my money back, I'll forget about it.")[6] Undaunted, he moved to Emmaus, a small town in Pennsylvania, and set up a small publishing venture that turned out pamphlets and magazines on how to achieve good health, using the knowledge he gained from trying to find alternative ways to combat his many ailments.

In 1941 Rodale experienced the health guru's *de rigueur* epiphany. In his case, it was reading one of Sir Albert Howard's books on the dangers of chemical fertilizers and the merits of manure and compost. The ideas, he later said, "hit me like a ton of bricks." Using the profits from a kind of thesaurus he published, he bought a farm outside of town and began experimenting with chemical-free farming. The next year he started a monthly magazine, *Organic Farming and Gardening*, to promote these methods. He sent out 10,000 free copies to farmers across the nation, but not one of them took out a subscription. They were profiting too much from following the wartime government's appeals to increase production with chemical fertilizers. Rodale had more success when he changed the title to *Organic Gardening and Farming*, something that struck a more favorable chord among the many city dwellers who were now cultivating Victory Gardens.[7]

While his ideas about farming derived from Howard, Rodale's views about how processed foods caused disease and how "natural" ones warded them off owed much to McCarrison. He was particularly struck by McCarrison's description of how the Hunzakut, as the people of the Hunza Valley were officially called, lived long, practically illness-free lives. In 1948 he published a book, *The Healthy Hunza*, that relied heavily on McCarrison's claims that their remarkable health and longevity came from "living on the unsophisticated foods of Nature."[8] The advertisement for Rodale's book in the *New York Times* read: "No cancer here . . . nor ulcers, dyspepsia or the degenerative diseases that trouble so many of us."[9]

Of course, Rodale had never been to the Hunza Valley nor met a Hunzakut, but by then there were other reports that seemed to confirm McCarrison's optimistic view. In particular, there was a 1938 book by an English doctor, G. T. Wrench, called *The Wheel of Life: The Sources of Long Life and Health among the Hunza*. It used McCarrison's ideas about natural foods to advocate a kind of "holistic" medicine. After being reprinted in America, it became, and remains, one of the basic texts of the health food movement.[10] When, in 1950, Rodale began publishing the monthly magazine *Prevention*, it regularly used the Hunza as examples of how eating natural foods could ward off the illnesses caused by the over-civilized diet.

At first, Rodale's message fell on quite deaf ears. American farmers were increasing production exponentially by dousing their fields with new chemical fertilizers and spraying their crops with DDT, the pesticide that was hailed as having won the war in the Pacific and whose inventor was awarded a Nobel Prize in 1948. For food processors, the 1950s were what I have called the "Golden Age of American Food Processing"—an era when the American media celebrated, rather than criticized, the chemicals and processes used in manufacturing foods. Thus, in 1950, when a congressional committee revealed that since 1938 food processors had

A reprint of the 1936 book that condemns food processing and attributes the supposed longevity and good health of the inhabitants of Pakistan's remote Hunza Valley to their diet of unprocessed "natural foods." The book helped inspire J. I. Rodale to launch the organic food movement in America.

introduced almost 850 new chemical additives into the food supply, the processors themselves began proudly issuing lists of the "new chemicals" they were using, pointing out that the Food and Drug Administration had approved them all as "Generally Recognized as Safe."

The congressional committee ending up proposing that in the future, instead of the FDA having to prove that new additives were unsafe, processors would have to prove they were safe. However, its chairman, James Delaney, never expressed any doubts that the vast majority of these new additives were safe and useful. Nor did the head of the FDA, Frank Dunlap (the man who betrayed Harvey Wiley in 1911, when he was "Old Borax's" deputy). He kept reaffirming that, unlike Wiley, he was not prejudiced against chemical additives per se. Only reluctantly did the committee's chief counsel (who initially refused, he said, "to put any nuts on the witness stand") allow J. I. Rodale to testify against the new chemical additives, and that was on the last day of hearings, when he was practically ignored.[11]

The committee ended up proposing that companies submit tests of future additives to the FDA. Yet even though this posed no threat at all to any of the chemical additives already in use, for the next five years Delaney and his bill were, in Delaney's words, "completely ignored." Then, in 1958 the National Cancer Institute released a study (that the FDA tried to have suppressed) saying that a number of the FDA-approved additives caused cancer in rats and might therefore do so in humans. This (along with an intense lobbying campaign among congressmen's wives by the film star Gloria Swanson, a disciple of Gayelord Hauser) suddenly spurred passage of Delaney's bill. Now, not only would food companies have to prove that new additives were safe for humans; they were also forbidden from using any additives that had been shown to cause cancer in animals.[12]

A year later this so-called Delaney clause set off a bombshell.[13] A few weeks before Thanksgiving, the secretary of health, education, and welfare announced that traces of a carcinogenic weed killer had been found on some of that year's cranberry crop. He consequently issued a warning against eating cranberries from Washington State and Oregon until they had been tested. Needless to say, that year few Americans had cranberry sauce with their turkeys.[14]

Then came revelations that a chemical used to promote hormone development in chickens caused cancer in other animals. This was followed by the news that one of the most commonly used food colorings caused cancer in rats.[15] The coup de grâce for the chemical Pollyannas came in 1962, when Rachel Carson's sensational book, *Silent Spring*, warned that pesticides were killing off birds and seeping into the human food supply. "For the first time in history," she said, "every human being is subjected to contact with dangerous chemicals from the moment of conception until death."[16] Surveys now showed that the dangers of chemical food additives were at the top of the list of consumer concerns.[17] Suddenly, the connection between the Hunzas' good health and their diet of natural foods began to seem plausible.

By then, as well, the mainstream media had begun taking notice of the Hunzas' longevity and crediting it to their simple diet. In 1957 C. L. Sulzberger, the *New York Times*' chief foreign affairs specialist, reported from northern Pakistan that "in Kashmiri Hunza generations live to an ancient age on a diet of dried apricots and powdered milk." That year Lowell Thomas, the renowned American movie journalist who had made

Lawrence of Arabia famous, featured the Hunza in *Search for Paradise*, one of the first feature-length movies made for widescreen Cinerama. As the huge screen filled with happy-looking Hunzakuts playing polo and sword-dancing, the soundtrack carried a cheery little song written by Thomas saying that in this bit of paradise, "nobody has too little, nobody has too much," and "you don't need pills for your ills."[18] The French-trained producer-director Zygmunt Sulistrowski made an English-language movie called *The Himalayan Shangri-la* extolling the Hunza diet as promoting eternal youth.[19] In 1959 Art Linkletter, the host of *People Are Funny*, one of the era's most popular television programs, sent a Des Moines, Iowa, optometrist to the Hunza Valley to test his theory that good health could be measured in the eyeballs. Upon his return, the optometrist told viewers that his eye exams showed they did indeed enjoy perfect health. He then wrote a book, *Hunza Land: The Fabulous Health and Youth Wonderland of the World*, extolling their diet. This joined many other books on the Hunzas' remarkable health, such as *Hunza: Adventures in a Land of Paradise* (1960); *Hunza: The Himalayan Shangri-la* (1962); *Hunza Health Secrets for Long Life and Happiness* (1964); *Hunza: Secrets of the World's Healthiest and Oldest Living People* (1968); and, some years later (my favorite title), *Hunza: The Land of Just Enough* (1974).[20]

These paeans to the happy, healthy people of Shangri-la described a practically inaccessible valley inhabited by from 30,000 to 50,000 people. They were descendants, it was said, of three soldiers in Alexander the Great's army who, with their Persian wives, either got lost or deserted in the Himalayas. Although in the past they had been brigands preying on the caravans of the Silk Road, they were now a peaceable, mainly agricultural people, who cultivated small plots of terraced land intensively, grazed some goats, and tended their apricot trees, to whose fruit many visitors ascribed extraordinarily healthful properties. They were practically self-sufficient and, being virtually outside a money economy, purchased hardly any processed foods. Their gracious ruler, the Mir, was a peaceful soul, who regularly gathered village elders in front of his modest home to settle disputes and issue benign edicts. These elders were reported to be remarkably alert and vigorous men, mainly in their eighties and nineties, with a significant number of centenarians. It was not uncommon for people to live to be 120 years old. Older Hunza men and women looked far younger than their age, and elderly men were said to lead active sexual lives well into their nineties. In the population at large, there was no sign of any illness. The children were plump and happy, the men had incredible physical stamina, and the women, Muslim but unveiled, radiated well-being. There were no hospitals and no jails, presumably because there was no need for either.

This kind of optimism was by no means confined to faddists. In 1959 President Eisenhower's cardiologist, Dr. Paul Dudley White, had a U.S. Air Force doctor visit the Hunza Valley to do electrocardiograms on twenty-five of its legendary old men and test their blood pressure and cholesterol levels. After examining the results, White ascribed the resulting "normal" readings to heredity, exercise, and their vegetarian, low-fat diets. "A man in really good shape can eat 3,000 apricots in a sitting," he reported.[21] In 1961 the *Journal of the American Medical Association* ran an editorial calling the Hunza Valley a harbinger of hope for America. It was highly possible, it said, that Hunzakut males lived to the ages of 120 and 140, "even though [they eat] little animal protein, few eggs, and no commercial vitamins. Their diet consists of whole grains, fresh fruit, fresh vegetables, goat's milk and cheese, rice and grape wine. The men are reported to be fertile in the ninth decade of life." (Characteristically, the journal's editors saw the hope for Americans as lying not in adopting the Hunza diet, but in modern medical science. They concluded, "If scientific advances continue to be productive, life expectancy in America should approach that of the fabled Hunzukuts by the end of this century and equal [it] in the 21st century.")[22]

Alas, the rosy picture of life in Shangri-la hardly corresponded to reality. In 1956 an American geologist who had spent almost a year in the Hunza Valley wrote that when he set up a small medical dispensary there, he was swamped with patients—over 5,000 of them—many of whom trudged there from long distances away. Ailments such as malaria, dysentery, trachoma, and worms were rife. There were many cases of scabrous sores, which he thought were malnutrition-related and treated with vitamin pills. Each spring, he noted, practically the entire society ran out of food and endured a period of famine until the first barley harvest. The Mir, who owned one-quarter of the best agricultural land, endured no such deprivation. On the contrary, he was a very hearty eater who suffered from stomach pains after practically every meal and came to the dispensary for milk of magnesia to relieve them.[23] Two years later Barbara Mons, an Englishwoman, reported from the Hunza Valley that although the people may have had great stamina, "that they possess a mysterious immunity to disease is not true." The one doctor who visited the area that year told her he had treated a large number of illnesses, including many hundreds of cases of dysentery and 426 of goiter, which is usually caused by a deficiency of iodine in the diet. There was no way to judge the claims to extraordinary longevity, she pointed out, because 99 percent of the population were illiterate (only the family of the Mir and a few close associates attended Shangri-la's one school). They therefore kept no records of births and had no clue as to when they were born.[24]

In 1960 the methodical Japanese sent a team from Kyoto University to examine the supposedly happy, healthy Hunza. Psychologists tried to assess their happiness with Rorschach tests, predictably coming to inconclusive conclusions. However, the doctor on the team again noted the rampant signs of poor health and malnutrition—goiter, conjunctivitis, rheumatism, and tuberculosis—as well as what seemed to be horrific levels of infant and child mortality, which are also signs of poor nutrition.[25] The year before, although he was a devout Muslim, the Mir had begged American visitors from a Protestant missionary hospital in Karachi to set up a mission hospital. "Three to four hundred boys, not to mention girls, died of whooping cough in Hunza this year," he told them. "Ninety per cent of my people have worms. Many of them have eye diseases, goiters, liver abscesses."[26]

But such reports hardly shook Hunzaphilia in America, for many of those enamored with the Hunza diet, and natural foods, rejected the kind of science and medicine that the Japanese and others brought to bear on the subject. To them, modern science and technology were the problem, not the solution. They had created modern food processing, which was denaturing the food supply, causing illness and death.[27]

In this respect, Hunzaphilia differed markedly from the lionizing of far-off peoples that accompanied the yogurt craze of the early 1900s. Yogurt owed much of its popularity to Metchnikoff's giving it the imprimatur of established science. Natural foods enthusiasts such as Rodale, on the other hand, wanted to hear about how the Hunza experience flew in the face of orthodox scientific wisdom. They ascribed the Hunzakuts' good health to their very ignorance of modern science and medicine and their instinctive adherence to the rules of nature. "The scientist," said Rodale in his book on the Hunza, "sits in his ivory tower, vesting himself with the cloak of omniscient authority. Nature is a menial serf, too lowly to obtrude on his attention." Another enthusiast said, "The people of Hunza are, from infancy to old age, continually attuned to nature." They enjoy extraordinary health and happiness because "they lead a life that nature intended and equipped man to lead. They eat the food that nature provided for man to use," a diet consisting of "natural, unprocessed, uncooked foods."[28]

Hunzaphiles thus described this healthy diet as having evolved from years of folk experience, not science. Rodale's *Encyclopedia of Organic Gardening* said that the "remarkable" Hunza apricot, which was "one of the secrets of the Hunza people's strength and health," had been "developed over 16 centuries." Another Hunzaphile wrote that since the Hunzas' few cows and goats did not produce enough fat for humans to eat, many years ago Hunza women "by instinct, or natural wisdom, found it in the most unlikely suspect—the seed of the apricot. And ever since, this knowledge has been passed down from mother to daughter." John Tobe's 1960 paean to them says: "Good health is a natural development of the way of life of the people of Hunza. It will remain theirs as long as they follow the same simple life that they have followed for almost 2000 years. . . . If [they] buy into civilization [there will be] a breakdown in their health and a shortage of their long life."[29] In other words, the Hunza diet was healthy because it avoided the depredations that modern science and industry made on our food.

Rodales and Radicals

In the late 1960s, this kind of neo-Romantic elevation of nature and folk wisdom over science and reason appealed mightily to young people swept up in the so-called counterculture. Soon long-haired people wearing beads and sandals were opening stores selling "natural foods" in practically every city and college town in the nation. The idea of avoiding foods processed by large companies also appealed to the political rebels of the New Left, who blamed giant corporations, and the scientists working for them, for the disastrous Vietnam War and most of the other things wrong with America. Although J. I. Rodale himself felt little affinity for "hippies" and the New Left, people in both groups found support for their views in the Rodale publications. Under his son, Robert, who gradually took over editorial direction of the company in the 1960s, the magazines regularly denounced large food processors, corporate farming, and conventional science and medicine while extolling the superior health of the darker-skinned poor people of the world who did not eat processed foods. Although at times they did give scientific-sounding explanations for the superiority of unprocessed foods, these usually reflected J. I. Rodale's vitamania, which emphasized that modern farming and processing deprived foods of their vitamins and minerals.[30] (The elder Rodale credited his being able to stand up to the "clobbering and insulting" of the early years to "plenty of vitamin B, the nerve vitamin," and said he took seventy vitamin and mineral supplements a day "as protection against pollution" and to restore vitamins lost in cooking.)[31]

At the same time, though, mainstream scientists began providing scientific backing to the Rodales' critiques. Rachel Carson's *Silent Spring* in particular led to a number of scientific studies warning of dangerous levels of mercury in freshwater fish and DDT residues in milk, fish, and other foods. By 1969 almost 60 percent of Americans surveyed said they thought that agricultural chemicals posed a danger to their health, even if used carefully.[32] Partly as a result, subscriptions to *Organic Gardening and Farming* jumped 40 percent, to 700,000, from 1970 to 1971.[33]

Meanwhile, young radicals were doing scientific research that supported other Rodale criticisms of the food supply. A number of them were recruited by the consumer advocate Ralph Nader, who in 1968 began denouncing the giant food "oligopolies" for ruining the nation's health. In 1969 he charged that "the silent violence of their harmful food products" was causing "erosion of the bodily processes, shortening of life, or sudden death." The culprits, he said, were the large food producers, who cared only about maximizing profits and minimizing costs "no matter what the nutritional, toxic, carcinogenic, or mutagenic impact may be on humans or their progeny."[34] That year 60 percent of Americans surveyed by the government said they thought that much of their food had been "so processed and refined that it had lost its value for health," and almost 50 percent agreed that the chemicals added to it "take away much of its value for health."[35] The next year a group of "Nader's Raiders" wrote *The Chemical Feast*, which blamed the supposedly deteriorating health of Americans on the chemicals used on and in their food. This was followed by Frances Lappé's best-seller *Diet for a Small Planet*, which called for eating natural, unprocessed foods that were "low on the food chain." Feminists climbed aboard, and the popular feminist health book *Our Bodies, Ourselves* repeated these warnings about the dangers that food additives posed to health.[36]

Natural foods also received a boost from the extraordinary popularity of Adelle Davis, who Nader's followers in the Center for Science in the Public Interest placed alongside Rodale in the "Nutritional Hall of Fame."[37] Americans bought close to 10 million copies of her books condemning white bread, sugar, and other "refined" foods. Long life and good health would come, she said, from drinking raw milk and eating homegrown vegetables, homemade granola, fruit, eggs, plain yogurt, brewer's yeast, blackstrap molasses, and plenty of vitamins. Confidence in her diet was somewhat shaken in 1974 when she died of cancer, at age seventy, especially since she had explicitly claimed it prevented that dread disease. However, shortly before her death, she said that when first told of the diagnosis, she too had reacted with shock and disbelief. "I thought it was for people who drink soft drinks, who eat white bread, who eat refined sugar, and so on," she said. However, she then traced the causes back to her student years, when she lived on "junk food."[38]

Meanwhile, in July 1972, J. I. Rodale had died at age seventy-two, despite his natural foods diet and avoidance of wheat and sugar (which he thought made people overly aggressive). Only weeks before, a piece on him in the *New York Times Magazine* had quoted him as saying, "I'm going to live to be 100, unless I'm run down by a sugar-crazed taxi driver."[39] The article had led to an invitation to be interviewed on the *Dick Cavett Show*, the most highbrow of the era's late-night television talk shows. The interview, which was taped a few hours before the show was to be aired, went very well. The short, goateed Rodale, whose looks reminded Cavett of Leon Trotsky (others said Sigmund Freud), displayed his hallmark ironic, self-deprecating humor, and Cavett made a mental note to have him back. Then, after Rodale shifted to the couch beside the interviewee's chair and Cavett began his conversation with the next guest, Rodale let out a snoring-like sound and slumped over, dead from a heart attack. The show was never broadcast, but the news of what had happened had such a tremendous impact that for years thereafter people would tell Cavett that they actually saw the show.[40]

Rodale Goes Mainstream

Unlike the untimely deaths of other health gurus, Rodale's passing hardly shook confidence in his ideas. In large part this was because organic farming and natural foods were already leaving behind the world of faddism and making inroads into the mainstream. Indeed, by the time of his death, *Prevention* magazine, with its natural foods message, was already selling 1 million copies a month.[41]

In the years that followed, changes in American political culture helped make middle-class Americans even more amenable to natural foods. Confidence in the kind of collective, government-sponsored solutions to problems that had characterized the Progressive Era, New Deal, and most recently President Lyndon Johnson's Great Society faded rapidly in the 1970s. The kind of faith that had accompanied the Meat Inspection Act of 1906 and the two Food and Drug acts now seemed quaint and naive. Once again, the competing tradition, which mistrusted government and held individuals responsible for creating and solving their own problems, came to the fore. Ironically, although these attitudes were soon associated with neo-conservatism, they also owed much to the long-haired rebels of the counterculture who, like their Romantic predecessors in the nineteenth century, saw salvation and good health as coming from individual dietary change. Although "hippies" no longer roamed the streets, their reverence for the "natural," including in foods, was taken up, along with their tastes in music and clothes, by many in the mainstream middle class. These soon became commercialized and, shorn of their original anti-establishment goals, were promoted by exactly the kind of large corporations they were originally designed to oppose.

The trajectory followed by the Rodale enterprise was not dissimilar. At first, Robert Rodale maintained his publications' anti-scientific-establishment posture. Commenting on his father's death, he praised his legacy of questioning people's faith in science. "You must look at nature and see what lessons she teaches," he wrote, "not just blindly use all the new chemicals that are given to us by white-coated technicians."[42] In May 1974 *Prevention* attributed Hunza longevity to the fact that, unlike Westerners, the foods in their diet were extremely high in "nitrolosides," a nutrient present in apricots that, under its other name, laetrile, was being touted by dubious practitioners of "alterative" medicine as a cure for cancer. The magazine also continued to delight in taking pot shots at orthodox medical advice about food. A piece on constipation, for example, told of how since the 1950s doctors had recommended first psychotherapy, then antibiotics, and then laxatives to cure constipation, whereas folk wisdom (that is, *Prevention*) had correctly advised eating natural foods with roughage, something that was still not endorsed by orthodox medicine.[43]

Increasingly, though, *Prevention* mustered conventional science in support of folk wisdom. That same year, in 1974, it cited the prominent scientists Jean Mayer and John Yudkin to support its attack on the FDA for saying that there was no proof that diet affected heart disease. "Hundreds" of studies, it said, "lead to the overwhelming conclusion that a diet which is as close as possible to that eaten by people living in close harmony with the soil . . . may well afford a valuable degree of protection against heart trouble." It also called the FDA's claim that modern food-processing techniques did not affect the nutritional quality of food a "ridiculous myth," citing scientific studies showing that heating and other forms of cooking reduced the vitamin content of many foods.[44]

That *Prevention* was able to refer to "hundreds" of articles on the diets of people "living close to the soil" was a sign of a profound change taking place in mainstream science: the kind of respect for the folk wisdom of traditional people that underlay Hunzaphilia was no longer confined to faddists and the counterculture. In 1973 *Scientific American, National Geographic,* and *Nutrition Today* all published articles by Alexander Leaf, a professor of gerontology at Harvard Medical School, on how living the simple life had endowed the Hunza as well as two other isolated groups, in the high Andes and Russian Georgia, with extraordinary longevity.[45] Leading medical and scientific researchers were now labeling heart disease, cancer, and other chronic illnesses as "diseases of civilization." Like the Rodales, they were now able to ascribe the apparent absence of these ailments among remote people to whatever dietary hobbyhorse they were riding. As a result, Robert Rodale was now able to do such things as back up his claim that fiber cured heart disease with a study of Ugandans in the well-respected *American Journal of Clinical Nutrition.*[46] Indeed, as we shall see in the next chapter, a study of heart disease and the diet among poor Cretan peasants was already leading to wholesale changes in American eating habits.

With a plethora of other premodern examples now at hand, references to the Hunza tailed off. Robert Rodale himself traveled to many Third World countries, reporting back on the connections he found between traditional agriculture, unprocessed foods, and healthy living.[47] In 1973 he had returned from China, where Mao Tse-tung's primitive (and, it was later revealed, disastrous) agricultural ideas still reigned, and told of how ancient farming techniques were producing healthful foods and creating a model for sustainable agriculture.[48] The following year he told of how American Indians had been making his favorite corn cakes—"a miracle food"—for thousands of years. They were the product, he said, of years of trial and error and, most important, common sense: the very thing that was the basis of Chinese herbal medicine.[49] In America, on the other hand, food production was "in the hands of corporate farmers and giant processing companies, [whose] standardized, chemically-treated products" were a "threat to health."[50]

However, the syndicated weekly column that Rodale wrote for such mainstream newspapers as the *Chicago Tribune* and the *Washington Post* from 1971 to 1975 was much less radical-sounding.[51] Indeed, many of his recommendations would soon become mainstream. He rejected the idea, prevalent in the 1960s, that vigorous exercise was bad for the heart and suggested regular exercise to avoid "heart pollution." (This led, in the 1980s, to his company founding the very profitable magazines *Runner's World* and *Men's Health*.) He was one of the first "locavores," advocating eating locally because of the amount of energy wasted in importing foods from afar. He also condemned excess food packaging for wasting energy and unnecessarily filling dump sites. He defied the large chorus condemning saturated fats by recommending free-range eggs as "a naturally good convenience food." Such sensible-sounding blends of concerns for health, taste, and the environment contributed significantly to mainstream acceptance of natural and organic foods. A 1978 profile of him in the *New York Times* portrayed him as much as a concerned environmentalist as a health food advocate and pointedly avoided labeling him a faddist.[52]

By then, the Rodales' battle to make Americans fearful of chemical food additives was well on its way to being won. Already in 1977 a majority of Americans surveyed thought that "natural foods," which they understood to mean foods that were free of chemical additives, were safer and more healthful than others. Food processors were now calling everything from potato chips and breakfast cereals to butter and beer "Natural," "Nature's Own," and "Nature Valley." Many claimed to be "additive free," without saying what was meant by additives. A study of food ads in women's magazines in 1977 showed that more than a quarter of the products were promoted as "natural," even though only 2 percent of the products could in any way be defined as "health foods."[53]

Good-bye, Hunza; Hello, Pete Rose

One way in which *Prevention*'s faddist bent had lingered on after J. I. Rodale's death was in a propensity to use dubious evidence to promote foods as cure-alls. A typical 1974 piece cited an obscure study with no scientific credentials to tell women that flavonoids—nutrients present in certain fruits—could be a substitute for hormone therapy and could cure varicose veins and hemorrhoids as well. (It is surely no coincidence that 70 percent of *Prevention* readers were women, whose average age was fifty-one.)[54] However, the growing supply of scientific literature on nutrition soon allowed the magazine to cherry-pick them and back up many of its nostrums with respectable journal references. The result was a steady drift away from science-bashing of the kind that originally underlay its Hunzaphilia. By 1990, when Robert Rodale died in a car accident in Russia, the balance of proof in *Prevention* had shifted decidedly toward mainstream science.[55]

Robert Rodale's death left the direction of Rodale Press in the hands of his wife and daughter. They continued Robert Rodale's drive to turn it into a diversified media empire specializing in health, fitness, and psychological well-being. In 2000, though, the company hit a bump in the road, and profits sagged. An outsider, Steve Murphy, was brought in from Disney Publishing to overhaul the enterprise. There followed the usual staff bloodletting. Over 120 old-line editors and employees were forced out, grumbling about the publisher's abandonment of its "mission." Within three years, though, the revamped company's finances had been turned around. In 2003 *Prevention*'s circulation was well over 3 million, and Rodale Press was on its way to becoming the nation's largest publisher of diet and health literature. That year it hit a jackpot with the weight-loss book *The South Beach Diet*, which sold 5 million copies in several months. Then came another best-seller: the autobiography of the disgraced baseball player Pete Rose.[56] By 2008 Rodale was said to be the largest independent publisher in the United States.

But the disgruntled ex-employees probably had a point. *The South Beach Diet* emphasized eating low-carbohydrate foods, not natural ones, and Pete Rose's book warned of the dangers of gambling, not processed foods.[57] In the first half of 2007, the company had ten books on the *New York Times* best-seller list—including *The Abs Diet for Women*, *LL Cool J's Platinum Workout*, and Al Gore's *An Inconvenient Truth*—hardly any of which reflected J. I. Rodale's original food concerns and nostrums.[58] That year it began publishing the wildly successful Eat This Not That series, which told readers which foods were (relatively) healthier in the kind of fast-food and chain restaurants that J. I. Rodale abhorred.[59] By then, it is doubtful that more than a handful of the over 1,000 people working for Rodale Press had even heard of the Hunza, whose example had so inspired its founder.

By then, as well, the original engine for the myth of the Hunza, J. I. Rodale's campaign for "natural foods," had been almost completely co-opted by the very large agribusinesses and corporations who had been its original targets.[60] Now, instead of using artificial flavorings to replace the tastes that are inevitably lost in processing, they used "natural flavors"—chemical compounds whose differences from artificial ones may be discernible to philosophers of science, but not to anyone else.[61] Practically every kind of foodstuff was now called "natural." (Tostitos said its corn chips were "Made with All Natural Oil," leading one to wonder what kind of oil was "unnatural.") By 2007 Americans were spending $13 billion a year on "natural foods," and the market was growing at 4 to 5 percent a year. That year, "all-natural" was the third most popular claim on new food products. The next year, it knocked off "low-fat" and "low-calorie" to become *the* most popular claim on new foods and beverages. Although the designation was essentially meaningless, it seemed to reassure consumers that the foods contained no hidden dangers.[62] In effect, the food processors were profiting by exploiting the very fears that their production methods had aroused.

Prevention adapted readily to this, blithely seeking advertisements from the very large processors that J. I. Rodale and his son had condemned so vigorously. They now carried ads for Kashi "natural" cereals and the "Back to Nature" line of cereals, "process cheese," and "macaroni and cheese dinner," neither of whose producers, Kellogg's and Kraft, identified themselves as the products' manufacturers.[63] All of this seemed to provide posthumous support for one of the more compelling concepts of the 1960s radicals: that American capitalism had a remarkable ability to defang opposition by co-opting it—that is, by adopting the rhetoric of radicals' programs while gutting their content.[64]

Equally disturbing for J. I. Rodale's legacy was that his other great crusade, for organic farming, seemed to be heading in the same direction. As the market for organic foods increased, the business of producing them became dominated by many of the same large corporations who dominated the regular food supply.[65] Initially, when the federal government stepped in to regulate use of the term "organic," there was hope that it would not become as meaningless as "natural." However, as pressure from these large producers to dilute the definition mounted, it seemed that the organic food industry might be headed in a direction that would turn Rodale's apparent success into a Pyrrhic victory.[66]

On the other hand, had J. I. Rodale lived, as he expected to, until the end of the twentieth century, he would at least have been able to enter "Hunza" into an Internet search engine and discover that his claims for their health and longevity were still alive and well in the world of alternative medicine and health foods.[67] And had he lived as long as many of the Hunza were said to and walked into the Whole Foods market in New York's Time Warner building in August 2007, he might have been heartened by what he saw: there stood a prominent display of "naturally grown" "Himalayan Harvest Organic Hunza Golden Raisins" and "Hunza Goji Berries," accompanied by a book called *Hunza: 15 Secrets of the World's Healthiest and Oldest Living People.*[68] But then again, perhaps not.

9
Lipophobia

Ancel Keys and the Mediterranean Dream

Juliet's lament "What's in a name?" does not apply only to vitamins. One wonders what would have happened if the fats in our food and bloodstreams had been called by their scientific name, lipids. Would avoiding the off-putting term "fat," with its connotation of obesity, have mitigated much of the fear of fats in food that has plagued Americans? Perhaps, but probably not. In retrospect, the wave of lipophobia—fear of dietary fat—that has swept over middle-class Americans since the 1950s was simply too powerful to overcome.[1]

As with many other fears, fear of dietary fat originated in alarm over a supposed epidemic—in this case, of coronary heart disease. Curiously, this notion that heart disease was rampant was actually an offshoot of the increase in Americans' longevity during the first half of the twentieth century. The major reason for this was a sharp decline in deaths from infectious diseases, particularly among infants and the young.[2] This led to a higher proportion of Americans dying of chronic diseases—ailments such as heart disease, cancer, and stroke, which tend to kill off people at an older age. Then, in the 1940s, changes in how causes of death were categorized on death certificates brought a spike in the proportion of those deaths attributed to heart disease.[3] As a result, by the end of the decade, medical experts were raising alarms about an epidemic of heart disease. By 1960 many of them thought the main culprit had been identified: dietary fat.

The best known advocate of this theory was Ancel Keys, a physiologist at the University of Minnesota's School of Public Health. A short, blunt, impatient man with a rather Napoleonic demeanor, Keys prided himself in his ability to endure tough physical challenges. In 1933, while at Harvard University's Fatigue Laboratory, he led an expedition to the Andes Mountains, where he spent ten days at an altitude of 20,000 feet—8,000 feet above the level at which lack of oxygen begins to cause people to pass out—testing its effects on his own blood. Seven years later, after he had moved to Minnesota, War Department officials came up with what he later called the "crazy idea" that his knowledge of the effects of altitude on humans would enable him to create field rations for paratroopers—even though the rations would be eaten on the ground, not in the air. The department was so impressed with what he produced—small waterproof packets containing a piece of hard sausage, biscuits, a chocolate bar, coffee, soup powder, chewing gum, and two cigarettes—that it issued what it called "K" rations (after Keys) to all combat troops. (Keys later admitted that K ration "became a synonym for awful food" and tried to disavow the "dubious" honor of having fathered them.)[4]

Then in 1944 Keys persuaded the government to fund another test of human limits. He recruited thirty-six conscientious objectors and put them on starvation diets for five months, meticulously recording their descent into physical and psychological hell.[5] After the war, Keys continued to study dietary deprivation, including conducting a study that lent support to Wilder's notions about the importance of thiamine.[6] However, he then changed direction. Like a number of other scientists in the nascent age of food abundance, instead of studying the health consequences of not having enough of certain foods, he began looking into the dangers of eating too much of them.

Keys had become curious about something that kept cropping up in the local newspapers: that many apparently healthy business executives were being struck down by sudden heart attacks. He managed to convince 286 middle-aged Minneapolis businessmen to submit to annual medical exams and interviews about their lifestyles and food habits. Forty years later, the most common risk factor for their heart attacks turned out to be smoking, but Keys was not looking at that. Instead, he tested their blood pressure and, most particularly, the level of cholesterol in their blood. Some scientists were already accusing this yellow, gummy substance of adhering to the walls of the arteries, creating thick plaques that slowed the flow of blood into the heart. They said that when these built up enough to block the flow of blood almost completely, coronary thrombosis resulted, often bringing death. Keys soon concluded that this buildup was indeed the main culprit in the businessmen's heart attacks. All that remained was to discover what was causing it.[7]

The physiologist Ancel Keys in 1958, when he began to achieve fame by condemning the saturated fats in the American diet as killers and advocating a "Mediterranean Diet" as an alternative. (University of Minnesota)

Keys's road to Damascus—where he experienced his scientific epiphany—was literally a road. In 1951 he was on sabbatical with his biochemist wife, Margaret, in Oxford, England, which was still plagued by postwar shortages of food and fuel. Winter there, he later recalled, was "dark and cold," and he and his wife shivered in an un-insulated, drafty house. Fortuitously, he recalled that at an international conference in Rome an Italian doctor had told him that in Naples, where he practiced, "heart disease is no problem" and had invited Keys to come and see for himself. So, in February 1952, Keys and his wife loaded some testing equipment into their little car and fled gloomy England, heading south for Naples.

Tales of people finding Utopias often begin, as did the story of Shangri-la, with them surviving storms, and sure enough, Key's story has the couple barely surviving a bitter snowstorm before reaching the safety of the tunnel under the Alps connecting Switzerland to Italy. When they finally emerged into the brilliant sunshine of Italy, they stopped at an open-air café, took off their winter clothes, and had their first cappuccino. There, Ancel Keys later recalled, "The air was mild, flowers were gay, birds were singing. . . . We felt warm all over, not only from the strong sun but also from the sense of the warmth of the people."[8]

In Naples, Keys was told that practically the only coronary patients in the city's hospitals were rich men in private hospitals. (No one seems to have told him that poorer Italians, especially in the south, clung—often with good reason—to the old notion that few patients emerged from hospitals alive and avoided them at all costs.) He then had a young doctor persuade some firemen from a nearby fire station to let him take their blood pressure, collect blood samples, and ask a few questions about their diets. Later that year, in Madrid, Keys took blood samples from some men in one of the city's working-class quarters, where heart disease was also said to be rare. Then he did the same with fifty well-off patients of his host, a prominent Spanish doctor, who had told him that heart disease was rife among them. Lo and behold, the Naples firemen and poor people in Madrid had significantly lower levels of cholesterol in their blood than the wealthy Madrileños, whose serum cholesterol levels were as high as those of the Minnesota businessmen. Although he did not do any formal dietary surveys, Keys concluded from observation and occasional questioning that wealthy Neapolitans and Spaniards ate much more fat than poor ones—as much as most Americans. To Keys, this seemed conclusive proof that high-fat diets caused high cholesterol in the blood, which in turn caused heart disease.[9]

At first, Keys's conclusions were greeted with considerable skepticism. His presentation at a major World Health Organization forum was a near disaster, as critics skewered the flimsy evidence upon which it was based.[10] Yet, like other compelling health nostrums, Keys's theory did have a sort of commonsensical appeal. Most important was that it made eminent sense to Dr. Paul Dudley White, the nation's most prominent cardiologist.[11]

White was also enamored with the quality of life in Southern Italy. In his case the affection dated back to 1929, when he spent four months on the idyllic isle of Capri writing the textbook that became regarded as "the Bible" on heart disease. Three years later, he and his wife took a bicycling trip through the beautiful Dordogne Valley in southern France, where they were smitten by the simple country food. In the spring of 1954, the two visited Keys and his wife in Naples, where they all enjoyed salubrious walks in the surrounding hills and picnic lunches of bread, cheese, fresh fruit, and wine. Clearly, it did not take much to convince White that the natives' simple diets held the key to their immunity from the epidemic of heart attacks.[12]

As president of the International Society of Cardiology, White was well-placed to spread the good word. That September he turned its international congress in Washington into a forum for research, with Keys's theory at center stage.[13] In the inaugural session, White and Keys presented their argument to 1,200 doctors packed into a hall that sat 800. They reported that in a hospital in Southern Italy, where high-fat diets were rare, only 2 percent of the deaths were due to coronary causes, whereas in White's hospital, Massachusetts General Hospital in Boston, it caused 20 percent of the deaths. In another paper, Keys argued against the competing notion that the heart disease epidemic was caused by being overweight. Dieting to lose weight would not reduce one's risk of heart disease, he said. "Heavyweights" were as common in nations with low-fat diets and little heart disease as in those with high-fat ones.[14] He told *Time* magazine that "being overweight isn't so much of a health problem as most people think. . . . Worrying about it can easily miss the real problem," which was heart and artery problems caused by high-serum cholesterol.[15]

The mainstream media now rallied to Keys's and White's side. The *New York Times* reported that specialists from around the world "agreed that high-fat diets, which are characteristic of rich nations, may be the scourge of Western civilization. The diets were linked with hardening and degeneration of the arteries." *Newsweek*'s report was more direct. "Fat's the Villain" was its headline, meaning dietary, not body, fat.[16]

Politics soon provided White and Keys with a golden chance to further publicize their message. Early in the morning of September 24, 1955, sixty-four-year-old President Dwight Eisenhower suffered a heart attack while staying at his mother-in-law's home in Denver, Colorado. With some fanfare, the air force flew White from Boston to the popular president's bedside. White caused quite a stir at the first of his twice-daily press conferences by saying the president had just had a normal bowel movement—something not usually discussed in public. Having caught the public's attention, he then used the regular press conferences to warn Americans about "the disease that had become the great American epidemic." After the president returned to the White House, White wrote a nationally syndicated newspaper article with his "reflections" on Eisenhower's heart attack. He said that although the presidency was a highly stressful position, the clogging of the arteries that led to the heart attack had little to do with stress. Nor was it connected with being overweight. In fact, the president weighed little more than when he was a West Point cadet. Rather, White said, it was likely related to consuming too much fat. "The brilliant work of such leaders as Ancel Keys" had shown this to be a major cause of coronary thrombosis. White had therefore put the hamburger-loving president on a low-fat diet. (There seems to have been no thought of telling him to quit smoking.)[17]

The day after White's piece appeared, *Time* magazine featured Keys in a cover story on "the nation's No. 1 killer," heart disease. "Atherosclerosis," it said, was "the real bugbear," attacking the coronary arteries with special frequency. The disease's prevalence in the United States and northern European countries had previously been ascribed, it said, to "racial origin, body build, smoking habits . . . the amount of physical activity . . . and, of course, the Gog and Magog of modern medicine: stress and strain." But Keys had knocked over these theories one by one. He said, "The popular picture of the coronary victim as a burly businessman, fat and soft from overeating and lack of exercise, who smokes and drinks too much because [of his stressful life] is a caricature." Such men often escaped coronary diseases, while other types fell victim to it. Nor were heredity and race involved. American Negroes and Italian Americans suffered the same rates of heart disease as other Americans, while Africans and Italians in their native lands were generally spared. As for smoking, Keys said, "Many peasant people smoke as many cigarettes as they can get, and often down to the last tarry fraction of an inch, without developing heart disease." With regard to obesity and being overweight, he would only grant that "gross obesity" might be "no more than some aggravating and accelerating influence." The only factors that counted, he said, were lack of physical exercise (White was becoming a fitness fanatic) and the percentage of fat in the diet. The "clincher" in this regard, he said, was to be found among Yemenite Jews. They had no coronary heart disease in Yemen but had begun to develop it since moving to Israel and adopting its high-fat diet. *Time* did acknowledge that whether cholesterol was indeed "the villain" was still "the hottest of all arguments," but the piece read as if Keys and White had already won the argument.[18]

White and Keys had little trouble getting funding to extend their studies to other nations. The two shared what Keys called "research adventures" with two more trips to Italy and visits to Japan and Greece. They also began employing the distinction that other scientists were now making between saturated fats—those of animal origin—which were said to raise blood cholesterol levels, and unsaturated ones, which did not. They returned from one trip announcing that they had checked 657 rural Cretans, ages forty-five to sixty-five, who got most of their fat from olive oil. Only two of them had suffered hearts attacks. They estimated that a similar group of American men, who ate large amounts of animal fats, would have had about sixty attacks. Keys then visited South Africa, Finland, Sweden, and Yugoslavia, where collaborators gave him evidence from blood samples and dietary questionnaires that seemed to confirm that coronary heart disease was related to high-fat diets and high levels of serum cholesterol. In eastern Finland, where heart disease rates were very high, Keys's recommendation that people give up such practices as spreading butter on thick slices of cheese earned him profuse praise from local public health authorities.[19]

Still, many scientists regarded Keys's evidence as scattered and insubstantial. In 1957, after he used World Health Organization statistics from six countries to support his argument, critics pointed out that if he had used the WHO data that was available from sixteen other countries, his conclusions would not have held up at all.[20] (Later, in 1977, a leading researcher would say of Keys's analysis, "The naïveté of such an interpretation of associated attributes is now a classroom demonstration.")[21] Even scientists who accepted the idea that high-fat diets produced high levels of cholesterol in the blood had a difficult time believing that this caused atherosclerosis. Others who were ready to admit that cholesterol might cause atherosclerosis were unconvinced that dietary cholesterol, as opposed to the much larger amount produced by the body itself, played much of a role. Audiences hearing the conflicting views at scientific conferences frequently left feeling that the matter was still very much up in the air.[22]

Ultimately, though, thanks in large part to White, Keys was able to gain government support for a far-ranging study designed to counter the objections to his theory. A key actor in this was Mary Lasker, a wealthy philanthropist who in 1948 began campaigning to have the federal government fund research to conquer cancer and heart disease. She managed to persuade key congressmen to resuscitate the National Institute of Health (NIH), which was languishing, and to have it create two special research institutes, one for cancer and another, the National Heart Institute, for heart disease. Paul Dudley White was recruited to participate in the latter, and over the next ten years he helped Lasker persuade Congress to steadily boost NIH appropriations for research on the heart disease "epidemic." After Eisenhower's heart attack, White was able to gain White House backing for a large increase in this funding.[23]

According to *Time* magazine, White also functioned as Keys's "chief fund raiser." This paid off handsomely in 1959, when the National Heart Institute agreed to help fund Keys's magnum opus, the *Seven Countries* study. Initially costing the then-extraordinary sum of $200,000 a year, this venture was a comparative study of serum cholesterol levels, diets, and rates of coronary heart disease and mortality among more than 12,000 men, ages forty to fifty-nine, in the United States, Italy, Finland, Yugoslavia, Greece, the Netherlands, and Japan. The American Heart Association, in which White was a major figure, provided additional funding.[24]

The *Seven Countries* study, though, was to run for twenty-five years, and Keys, who was already fifty-five, was not about to sit on the sidelines until it was completed. He continued his drumbeat of warnings against the dangers of dietary cholesterol, citing his studies in Italy and Crete as well as ones by collaborators in other parts of the world. Later he would confess that there were "mistakes" in these studies, but at the time he admitted to no such doubts, saying that they provided solid evidence to support his recommendations.[25] He and his wife then wrote a diet book, *Eat Well and Stay Well*, to teach Americans how to reduce their cholesterol intake by cooking the kind of healthy food enjoyed by Mediterranean people like their neighbors in Naples.[26]

Paul Dudley White's preface to the book seemed to presage some tough dietary changes. It told of how much he had enjoyed eating on the hardscrabble Italian island of Sardinia, even though it meant reducing his fat intake "from our customary 45 percent of calories to about 20 percent." But the book settled for recommending that 30 percent of calories come from fat, and prescribed dietary changes that were distinctly moderate.[27]

The "chief reason" for the book, said Keys, was to wage war against cholesterol, and an initial chapter told of recently discovered differences in fats in a simple, straightforward fashion: "Hard" fats—such as butter, lard, cheese, and lard-based margarine—were "saturated" and caused high levels of cholesterol in the blood and "probably" deposits in the coronary arteries. On the other hand, fats that were liquid at room temperatures, such as corn and cottonseed oil, were "unsaturated" and were "to be favored." There was no mention of the notion, which was already circulating, that there were two kinds of saturated fats: some that encouraged the body to produce "high -density" cholesterol molecules that did not seem to contribute to heart disease and others that encouraged production of the "fat and flabby" molecules—later to be called low-density lipoproteins (LDL)—that were the villains.[28]

This simple view of a "saturated/unsaturated fat" dichotomy allowed the Keyses to suggest a diet that was very easy to follow. Many of the recipes simply substituted skim milk for whole milk and vegetable oil for butter and lard. Because Keys thought that the "preformed" cholesterol in eggs and organ meats was not absorbed into the bloodstream, he saw no problem in eating three or four eggs a day. Nor did the book advise giving up red and fatty meats. There were recipes for cholesterol-rich calves' liver and for round steak stuffed with bacon. The dark meat and skin of chicken, objects of fear and loathing by later lipophobes, were not scorned. Indeed, not only did a recipe for "arroz con pollo" call for a whole chicken with its skin; it also included an entire Polish sausage.[29]

Today when we are inundated with foreign recipes striving for "authenticity," *Eat Well and Stay Well* seems to fall ludicrously short in teaching how to cook and eat the Mediterranean way. For example, there are only two recipes for pasta, and little indication that Margaret Keys had mastered standard Southern Italian ways of preparing it. Recipes with Spanish, Italian, and French names bear only a passing acquaintance with the originals. The recipe for "'Creamy' French Béchamel Sauce," for example, called for flour and MSG to be sautéed in oil and then blended with meat stock and skim milk.[30]

Yet the fact that the book was not a radical departure, either in dieting or culinary terms, undoubtedly helped it become an instant success. Its publisher, Doubleday, launched it with advertisements bearing the bold-faced headline: **"Will you commit suicide this year?"** It then warned, "Nearly 500,000 Americans will—unintentionally, unwillingly and needlessly—and you may well be among them." The Medicine section of *Time* magazine praised the book, saying, "Doctors still differ about many of the details of the relationship between a high-fat diet and the high death rate from coronary heart disease in the U.S., but more and more are coming to a practical conclusion: cut down on the fats without waiting for all of the facts." The book, it said, showed housewives how to follow doctors' advice to switch from butter and lard to vegetable oil, making lower-fat, cholesterol-reducing diets "practical in the average home."[31]

Doubleday then rushed out advertisements saying, "Here is the book you read about in the medical section of Time Magazine [that] makes clear the vital relationship between good health and good eating." It quoted Paul Dudley White as calling it "a happy blending of the scientific aspects of nutrition, the hazards of overnutrition, and the pleasures of the table." Within two weeks of its publication, the book was one of the top ten on the *New York Times* best-seller list. Six months later it was still selling well, and Doubleday was able to take out full-page ads filled with fulsome praise from doctors around the world. "Never before have so many doctors endorsed a diet book," it said.[32] Thanks to the book's royalties, Keys and his wife were able to buy a piece of the Mediterranean dream for themselves—a second home in the hills overlooking Naples.[33]

In late December 1960, Keys received another boost when the American Heart Association had him draw up a statement that said reducing the amount of saturated fat in the diet was "a possible means of preventing atherosclerosis and decreasing the risk of heart attacks and strokes." It called for reducing consumption of foods with high levels of saturated fats, such as dairy products and meat, and replacing them with foods such as vegetable oils that had low levels of saturated fat and high levels of polyunsaturated fats, which did not raise blood cholesterol. To support this, Keys cited studies showing that people in countries who had low-fat diets and lower serum cholesterol suffered fewer heart attacks than Americans. The AHA did force Keys to add that "as yet there is no final proof" of causality, but press reports on the statement quoted a former AHA president as saying that nine out of ten doctors were already "going ahead on the assumption that there is a relationship and that it's a good idea to trim cholesterol levels."[34]

Within hours, giant commercial interests were rolling out their artillery. Vegetable oil producers began vying over whose product was better at preventing heart attacks. Wesson Oil's full-page ads quoted the AHA statement that substituting polyunsaturated fats for saturated fats reduced blood cholesterol and "hence the risk of heart attacks," and said, "poly-unsaturated Wesson is unsurpassed by any leading oil in its ability to reduce blood cholesterol."[35] The National Dairy Council, thrust on the defensive, responded by calling the idea that replacing saturated fats with unsaturated ones would lessen one's chances of getting heart disease "clearly unproved" and possibly "dangerous to health."[36] This hardly impressed Wall Street, which drove up the share price of Corn Products, producers of Mazola oil, by over 5 percent.[37]

A week after the AHA statement, the lipophobe camp received another boost when *Time* magazine did a cover story on Keys. Entitled "Fat of the Land," it called him "the man most firmly at grips with the problem . . . of diet and health." He was conducting "a $200,000 a year experiment on diet" that spanned seven nations and was still growing. He had already "logged 500,000 miles, suffered indescribable digestive indignities [and] collected physiological data on the health and eating habits of 10,000 individuals, from Bantu tribesmen to Italian contadini." His conclusion: "The main culprit" in causing "the nation's No. 1 killer: coronary artery disease . . . was saturated fat in the diet, which raised levels of cholesterol in the blood. . . . He regards the cause-and-effect relationship between cholesterol and heart disease as proved."

The article then listed a number of foods, now including eggs, that contributed to heart attacks. High among these deleterious foods were what Keys called "carving meat"—steaks, chops, and roasts. Yet here again, as in the diet book, Keys's dietary recommendations were hardly extreme. He said that he and his wife enjoyed these meats three times a week and drank reduced-fat 2 percent milk, rather than skim milk. His recommendations were: "Eat less fat meat, fewer eggs and dairy products. Spend more time on chicken, calves' liver, Canadian bacon, Italian food, Chinese food, supplemented by fresh fruits, vegetables, and casseroles." He also emphasized the importance of enjoying eating. He and his wife had cocktails before dinner and ate slowly, with candles on the table and Brahms playing in the background.

But Keys was not at all restrained regarding the correctness of his theories. He said that the note in the AHA statement saying that "as yet there is no final proof" was inserted against his will. He himself was fully convinced that saturated fat in the diet caused heart disease.[38]

Yet Keys must have known that there could never be a "final proof" from studies such as his. The best that could be expected from such "prospective" epidemiological studies was a high association between disease and certain "risk factors." Even experts who agreed with Keys added caveats to that effect. Nevertheless, the public at large was easily convinced that science had rendered a resounding guilty verdict in the case against saturated fat. From 1959 to 1961, per capita consumption of milk and milk products declined by over 10 percent, while that of butter dropped by over 20 percent. In 1962 the dairy industry was thrown into a panic over a study by a Madison Avenue "motivational researcher" who told them that "belief in the dangers of cholesterol had taken a firm hold" on the public. People who had previously thought of dairy products as embodying health and vitality now associated them with cholesterol and heart ailments.[39] Whole milk, the "miracle food," was swiftly being transformed into a killer.

The Successful Man's Disease

Why were middle-class Americans so receptive to Keys's message? Part of its appeal was that it portrayed heart disease as a consequence of America's economic success. As we have seen, in the late nineteenth century, better-off Americans were attracted to the idea that dyspepsia and neurasthenia were by-products of the fast pace of life in America's booming business civilization. Now, with their country embroiled in the Cold War with the Soviet Union, Americans believed that the superiority of their system was exemplified by their affluence. Social scientists now saw practically all aspects of American society as products of affluence, including some negative ones.[40] In his 1958 book *The Affluent Society*, the economist John Kenneth Galbraith said that although private affluence had led to the disappearance of all but a few "pockets of poverty," there was persisting poverty in public services and institutions. Keys, who admired Galbraith, struck a similar chord, portraying heart disease as "a disease of affluence"—an unfortunate side effect of America's economic success. In other countries, he said, heart disease afflicted mainly rich men, who were the only ones who could afford fatty foods. Heart disease was much more widespread in the United States, he said, because practically everyone there could afford a rich man's diet.[41]

From the outset, the people raising alarms over the coronary plague had exploited this presumed association of heart disease with successful men. In 1947 the American Heart Association kicked off its first national fund-raising drive by warning that coronary thrombosis struck doctors, lawyers, and business executives much more frequently, and at a younger age, than it did manual laborers and farmers. (It provided no data to support this questionable claim.) The nationally syndicated medical columnist Dr. Peter Steincrohn, author of a book on heart problems, called men with heart disease "the successful failures . . . who accomplish much for their generation, but, like the busy bees, die off sooner than the rest of the hive."[42] This theme was repeated endlessly over the next ten years, culminating with *Time*, in its 1955 piece on Keys, calling it "a disease of successful civilization and high living. . . . The image of the tycoon who, at age 50, has attained money, success, a yacht and a coronary thrombosis is almost part of American folklore."[43]

Other contenders for responsibility for the "coronary plague" were also regarded as products of America's affluent business civilization: stress, obesity, sugar, and smoking, particularly the first of these. All Keys did was to change the cause of the busy bees' early demise from stress to "a rich man's diet."[44] This went along with a change of heart among nutritionists. They were beginning to think that their previous advice that "excess is preferable to limitation" was responsible for the obesity, atherosclerosis, and diabetes that now seemed prevalent among "the upper income groups."[45]

These fears of the consequences of over-indulgence were grounded in the perception that Americans were surrounded by an incredibly bountiful food supply. By the mid-1950s, the nation's farmers were producing so much food that they had to be paid not to produce it. By 1961 the government was trying to figure out how to dispose of 100 million tons of surplus farm products. Supermarkets overflowed with foods that had previously been unavailable or unaffordable to most people. One of the attractions of the major rival to Keys's theory, that being overweight was the cause of the heart disease epidemic, was that it faulted those who could not resist the temptations of abundant food. (Keys himself said he found obese people "disgusting" but argued that obesity might be implicated with heart disease only to the extent that "fat people tend to eat more fats.")[46] Keys played on this revulsion against the loss of self-restraint as well. Prosperity, he said, had allowed all Americans to "attain the food heaven of the teenager—unlimited fat steaks, French fries, and triple dips of ice cream."[47] Later, in 1975, Keys would use this appeal to restraint and simplicity when he repackaged his recommendations into the "Mediterranean Diet."[48]

However, as is so often the case, Keys's theory also triumphed because there was so much money to be made from it. During the 1950s, American food companies had become obsessed by what they called the "fixed stomach": That is, the idea that Americans were physically incapable of eating larger quantities of food than they already were. Some sought to increase profits by adding value to their products with new ways of processing and packaging. Others tried wresting a larger share of the static market away from similar foods. Among the most zealous at the latter were vegetable oil producers, who energetically stoked fear of saturated fats to sell their products as replacements for butter and lard.

They did this in the face of some early obstacles. Soon after the margarine manufacturers figured out how to replace the lard in their products with vegetable oil, it was discovered that the hydrogenation process they used to harden the oil transformed its unsaturated fats into the bogeyman, saturated fats. The result left margarine with as much cholesterol as butter. However, they quickly adapted, and in late 1960 began mixing some liquid oil into their margarines, producing a "partially hydrogenized" product with considerably less saturated fat than butter.[49] From then on, they had the butter producers on the run, firmly establishing their products as healthful heart-savers. Although some scientists warned of the possible dangers of the trans fats created by the process, Keys led the way in dismissing these concerns as groundless.[50]

By the end of 1961, then, Keys was riding very high indeed. His basic ideas about the dangers of cholesterol-laden foods were already part of the conventional wisdom. The media hailed him as the nation's top expert on food and health; the University of Minnesota lauded him as one of its star researchers; he traveled the world collecting data and advising local authorities on their nations' diets. But many scientists still remained skeptical, including some who were put off by his contemptuous treatment of those who disagreed with him. (He was, said one of them, "pretty ruthless" and hardly apt to win any "Mr. Congeniality" awards.)[51] Moreover, although a number of powerful interests were now disseminating his ideas, others were challenging them. The next decades would see a host of interest groups—scientific, commercial, charitable, political, and professional—weigh in to the debate.

10

Creating a National Eating Disorder

The Cholesterol Wars

Keys's scientific opponents did not fade gently into the night. Instead, for much of the 1960s, they fought back doggedly, provoking what at the time was politely called the "cholesterol controversy" but later was more aptly called the "cholesterol wars."[1] This revolved around two questions: First, do high levels of cholesterol in the blood cause heart attacks? Second, if this is true, can the high cholesterol levels be diminished by a low-fat diet? By 1970, although much of the medical research establishment was on board with the first proposition—that high blood cholesterol caused heart disease—proponents of the second proposition—that it could be significantly lowered by diet—were finding conclusive proof elusive.[2]

However, as far as the general public was concerned, both questions had been answered decisively in the Keysians' favor. Indeed, they were assured of this by the nation's most prominent nutritionists. In his nationally syndicated nutrition column, Frederick Stare, head of Harvard's Department of Nutrition, advised that heart disease could be prevented by eating fewer eggs and more margarine, vegetable shortening, and vegetable oil.[3] In his syndicated column, Jean Mayer of Tufts University, the nation's best-known nutritional scientist, warned that low-carbohydrate diets led to increased fat consumption, which was "the equivalent to mass murder."[4]

Equally important in the triumph of lipophobia was a new force in creating food scares: nonprofit health advocacy groups. Often begun by well-meaning people seeking to raise money to cure diseases, they could easily mutate into slick machines staffed by professional fund-raisers whose hefty salaries depended on alarming the public about the dangers posed by their particular illness.

This is exactly what happened to the American Heart Association (AHA). Originally formed in the 1920s by heart specialists to exchange ideas about their field, by 1945 it was raising a modest $100,000 a year to subsidize conferences and fund some research. Meanwhile, the March of Dimes, founded in 1938 to combat polio, was collecting $20 million annually. This could not have been far from the minds of the new leaders who took over the AHA and set out to arouse public concern about the "coronary plague." They hired Rome Betts, a former fund-raiser for the American Bible Society, to create a professional fund-raising apparatus. He recruited a star-studded cast of laypeople, including the Hollywood movie mogul Sam Goldwyn and the author Clare Booth Luce, wife of the powerful *Time-Life* publisher Henry Luce, to sit on a new board of governors. In the ensuing years, they helped recruit a host of celebrities to participate in the AHA's annual "Heart Week" drives, which raised money to fight "the greatest epidemic in the twentieth century," "the nation's most serious medical and public health problem."[5] The organization also benefited from President Eisenhower's heart attack and a campaign to send him "get well" letters accompanied by donations to the AHA.[6]

At the outset, those running the reconstituted organization realized that successful fund-raising would have to involve more than just pleas to support research. Only about one-quarter of its budget went for that, and there was little hope that there would be much to show for it in the near future. Instead, its new president said its goal was educating the public about "the significance of blood pressure, infections, overweight, rheumatic fever and other factors which contribute to various forms of heart disease." It especially targeted businessmen, sending specialists to speak at their meetings. (A Cleveland cardiologist told one group that they could head off high blood pressure, "the No. 1 killer of the average business executive," by taking naps at noon.) It tried to combat prejudice against businessmen who returned to work after heart attacks, something it said was just as important as supporting research. In 1954 Vice President Richard Nixon, speaking to the international Congress of Cardiologists, praised their success at this. He told of how eight months earlier the brilliant forty-four-year-old general who was President Eisenhower's liaison with the Pentagon had been stricken with heart disease and had now returned to work "as good as new." (Unfortunately, five days later the poor man succumbed to a fatal heart attack.)[7]

However, the philanthropist Mary Lasker, who was a major contributor to the American Heart Association, was interested mainly in funding research. This dovetailed with the interest that Paul Dudley White, a leading figure in the organization, was taking in Ancel Keys's research. As a result, in the mid-1950s support for research began to outdistance the organization's public health efforts.[8] The *New York Times*, whose controlling Sulzberger family were major contributors to the AHA, helped drum up support for this. The paper's science reporter Howard Rusk wrote enthusiastically about Keys's and White's work on the "coronary plague" and received one of the annual awards that the AHA began giving to journalists for coverage of heart research. In 1959 the AHA's president attributed his organization's success in raising a record $25 million to growing public awareness of the need for research.[9]

The association's success in arousing support for research soon proved to be a double-edged sword, for it also helped stimulate a spectacular increase in government funding for heart research at the National Heart Institute (NHI), the heart research arm of the National Institute of Health (NIH). In 1948 the two organizations each had budgets of around $1.5 million. In late 1961 AHA leaders were nonplussed by the fact that although their budget had risen to $26 million, the heart institute's now totaled $88 million. Worse, the NHI's budget was slated to rise to an astounding $132 million the next year. The AHA leadership now began worrying about being "squeezed out" by the federal government: that is, they feared that people would think that because their tax dollars were going for research into heart disease, there was no more need to contribute to the heart association.[10] If the AHA hoped to maintain its high profile and justify its money-raising efforts, it would have to return to emphasizing public health. This meant recommending lifestyle changes, something for which Keys's ideas provided the perfect springboard.

As we have seen, the AHA took a large step in this direction in 1960, when it backed Keys's theory that lowering dietary fat reduced the risk of heart attacks. However, it had only recommended low-fat diets for people with a greater than normal risk for heart attacks. In June 1964 it threw off these restraints and warned all Americans to reduce their total fat intake and to substitute vegetable oils containing polyunsaturated fats for saturated animal fats. It admitted that there was no proof that this would lower the risk of heart disease, but its spokeswoman said that heart disease had become "such a pressing public health problem" that it "just can't be left until the last 'i' is dotted and last 't' is crossed."[11]

Yet it was already apparent that dotting the i's and crossing the t's was going to be a lot more difficult than Keys had anticipated. Even as the *Seven Countries* study was getting under way, there were serious doubts about whether this kind of study could provide convincing evidence that dietary cholesterol caused heart disease. The obstacles to accurately measuring and comparing the food intakes and health outcomes of such disparate groups of people were simply insurmountable. Only "prospective" studies, using strictly supervised control groups who were fed on different diets and followed over many years, could hope to achieve this. However, a large NIH-funded study aiming to do this foundered, mainly because only about one-quarter of the men on low-fat diets could stick to them for more than a year.[12]

Pots of Gold at the End of the Yellow Fat Trail

Ultimately, no one ever succeeded in crossing those t's and dotting those i's. However, the public did not need this kind of "slam dunk" linking diet and heart disease to be convinced. One reason may have been the stunning impact of the 1964 Surgeon General's report on smoking, which used epidemiological evidence to make a convincing case that smoking caused lung cancer. Although the lipophobes could not present evidence anywhere near as compelling, their campaign, based on similar-sounding evidence, benefited mightily from this apparent triumph of epidemiology. Another shot in the arm for Keys's theory came from a more unlikely source, the American Medical Association (AMA). At first, the proposals to cure heart disease through diet seemed, like vitamania, to represent yet another threat to doctors' quasi-monopoly over treating illness. But by 1960 large numbers of patients, and their doctors, were becoming convinced that reducing dietary cholesterol would head off heart attacks. This propelled the AMA into trying to assert control over the new therapy. In 1960 it issued a report agreeing with that year's American Heart Association statement that those at elevated risk for heart disease should eat less fat and replace saturated fats with polyunsaturated ones, but it warned that this should only be done "under medical supervision."[13]

Vegetable oil and margarine producers quickly exploited the AMA's consequent recommendation to replace butter and lard with vegetable oil. Nucoa and Fleischmann's margarines each claimed to be the highest in healthy polyunsaturated fats. General Mills claimed that its "Saff-o-Life" safflower oil (an oil that until recently had been used mainly in varnish, solvents, and linoleum) had more polyunsaturated fats than any other oil. The major producer of safflower oil rushed out its own "Saffola" oil, whose ads began, "Read what the A.M.A. says about the regulation of dietary fat."[14] Other food producers joined in, with claims such as "no cholesterol," "low cholesterol," and "unsaturated." Soon polyunsaturated oil was being touted as a health food in its own right. The Harvard nutritionist Frederick Stare advised swallowing three tablespoons of it each day as a "medication." Jean Mayer, his counterpart at Tufts, said that consuming one cup of corn oil a day would prevent heart disease, but that it did not have to be in raw form.[15]

This blitz—particularly ads such as Saffola's saying that it recommended using vegetable oil to cut heart disease—alarmed the AMA. It tried to scotch this notion that people could prevent heart disease without the help of their doctors by issuing a statement saying, "The antifat, anticholesterol fad is not just foolish and futile; it carries some risk. . . . Dieters who believe they can cut down their blood cholesterol without medical supervision are in for a rude awakening. It can't be done. It could even be dangerous to try."[16] Yet when people did consult their doctors, they were almost inevitably told to go on low-fat diets. As a leading medical scientist later wrote, "Physicians were overwhelmed by this assault, both from their waiting rooms and their professional journals. A low fat, low cholesterol diet became as automatic in their treatment advice as a polite goodbye."[17]

"Should an Eight-Year-Old Worry about Cholesterol?"

The AMA's opposition to selling food for therapeutic purposes once again lined it up with the Food and Drug Administration, its erstwhile ally in the battle against vitamin supplements. In late 1959, the FDA had warned food producers that any claims that unsaturated fats would reduce blood cholesterol were "false and misleading." It had been shown, said its commissioner, that the human body produced its own cholesterol and was "affected very little by the amount present in our foods."[18]

But this did little to stifle the dubious claims. Finally, in September 1964, the FDA seized a shipment of Nabisco Shredded Wheat on the grounds that the packages carried "false health claims," namely that eating a bowl of it each morning would lower blood cholesterol and prevent heart disease and stroke.[19] For the rest of the decade, the agency repeatedly said that there was no proof of a relationship between dietary cholesterol and heart disease, and warned food processors against claiming that the polyunsaturated fats in their products might prevent heart disease.[20]

However, the Food and Drug Administration only controlled drug advertising. It was the Federal Trade Commission (FTC) that oversaw food advertising, and under its benign gaze, food processors ran all kinds of ads strongly implying that their products would prevent heart attacks. The secretary of the American Medical Association's Council on Foods complained that "we are all tired now of the unending advertisements for oils and margarines that promise to clean one's arteries in much the same way a drain cleaner works," but the AMA's journal itself carried such advertisements. The one for Fleischmann's margarine featured a photograph of a boy blowing out eight birthday candles. In medical journals, the caption said, "Is there a heart attack in his future?" and the text below recommended low-saturated-fat diets to prevent heart disease "*for people of all age groups*." In popular magazines, the company's ads advising parents to feed their children its margarine carried the same picture with the caption, "Should an 8-year-old worry about cholesterol?"[21]

In 1971, when the Federal Trade Commission finally stepped in to tell Fleischmann's to "tone down" its ads, it was hard to tell whether to laugh or cry. The commission forbade the company from directly claiming that its margarine prevented heart disease, but it did allow it to say that it "can be used as part of a diet to reduce serum cholesterol which can contribute to the prevention and mitigation of heart and artery disease." In justifying this, the FTC said that although these claims "have not been established by competent and reliable scientific evidence," heart disease was such a serious problem that it was good to "acquaint the consumer . . . with some of the steps recommended for its avoidance." Moreover, the order applied only to Fleischmann's and did not stop others from advertising, as did Saffola, that their foods "will do your heart good." Nor did it stop Fleischmann's from giving mothers a little booklet about its margarine called "The Prevention of Heart Disease Begins in Childhood."[22]

In fudging the issue, the FTC was simply going along with the American Heart Association, which routinely blamed dietary cholesterol for heart attacks while admitting, in the fine print, that there was no proof of this. In 1971, after the FTC reported that a majority of experts it consulted supported the diet-heart hypothesis, one of the dissenting scientists said bitterly, "The dietary dogma was a money-maker for segments of the food industry, a fund raiser for the Heart Association and busy work for thousands of fat chemists. . . . To be a dissenter is to be unfunded because the peer-review system rewards conformity and excludes criticism."[23]

The next year saw the AMA return to the lipophobe fold. This may not have been unconnected with the new role that had emerged for doctors—doing regular cholesterol checks. It now warned that most Americans had elevated cholesterol levels and that doctors should begin checking their cholesterol levels in early adulthood. Those found to be "at risk" should cut down on saturated fats—under their doctor's supervision, of course.[24]

Finally, in January 1973, the FDA caved in. It began allowing food labels to carry their cholesterol and saturated fat content and to say that eating foods low in cholesterol and saturated fats and high in polyunsaturated fats would help lower cholesterol levels.[25] Although they could still not say directly that this would prevent heart attacks, there was no need to do so. As Edward Pinckney, a disgusted preventative medicine physician, wrote:

> The consumer's understandable fear of heart disease and impending death is being exploited by certain health groups as well as by an industry whose profits have more than doubled as a direct result of its implied promise that heart disease can be forestalled through use of its products. The very word "polyunsaturated" has become synonymous with protection against heart disease just as "cholesterol" and "saturated fat" have been made to intimate doom.

Yet, he said, there was no proof at all that cholesterol in the blood caused heart disease and especially that dietary change would make any difference.[26]

However, naysayers like him were drowned out by the lipophobic chorus. The AHA soon raised the ante by telling all Americans to double their consumption of polyunsaturated fats and reduce their meat consumption by one-third.[27]

John Yudkin and the Challenge of Sucrophobia

In the late 1960s, with vegetable oil makers avidly funding anti-cholesterol research, the National Dairy Council tried to fight back with scientists of its own.[28] It recruited a professional staff of 320 people and gave them $14 million a year for "educational/scientific" work to demonstrate their products' healthful qualities.[29] But the most serious challenge to the Keysians seemed to come from a different direction. Hard on the heels of Keys's theory came a rival one that said carbohydrates, especially sugar, were a primary cause of heart disease.

Its main proponent was the English scientist John Yudkin, a man who, except for being short, had little in common with Keys. The son of impoverished Russian-Jewish immigrants to London, Yudkin was able to earn a PhD in biochemistry from Cambridge followed by an MD thanks to a series of scholarships. During World War II, he served as an army physician in West Africa, where he did some highly regarded vitamin research. He then became a professor of nutrition at the University of London. In 1958 he came to national prominence with a best-selling book, called *This Slimming Business*, that attacked "yo-yo dieting" and said that the best way to lose weight was to cut back on carbohydrates, especially sugar.[30]

The year before, Yudkin had joined those who criticized Keys for using WHO data from only six countries to support his dietary fat/heart disease theory. He pointed out that had Keys used the data that was available from ten additional countries, it would have shown that there was a better relationship between coronary mortality and consumption of sugar than of fat. The best relationship of all, though, was between the growing number of coronary deaths in the United Kingdom and the increase in radio and television sets. Of course, Yudkin made this last point to show that a close association between events does not necessarily mean cause and effect. However, he added that the rising number of TV sets did reflect growing affluence, and that the rise in coronary heart disease was likely connected with the things that went along with it—namely, increased smoking, obesity, "sedentariness," and, he strongly suspected, sugar consumption.[31]

Yudkin then tried to get firmer backing for his theory by comparing the sugar intake of men hospitalized with coronary heart disease with that of a number of healthy men. Here he found that first-time heart attack sufferers consumed twice as much sugar as those who had not suffered heart attacks. (No one, he pointed out, had ever shown that there was any difference in fat consumption between people with and without heart disease.) He then conducted laboratory experiments on rats that seemed to show that eating sugar greatly increased their levels of triglycerides, which were said to contribute to heart disease.[32]

Yudkin's fame, along with his quick wit and engaging personality, ensured wide coverage of his ideas.[33] The lipophobes, supported by Britain's powerful sugar industry, tried deriding him in the media. (He successfully sued one of them for calling his work "science fiction.") Ultimately, though, they brought him down by using their influence with research granting agencies to drain him of research funding. He did manage to get some funding from the dairy industry, but not enough.

In 1970 he was persuaded to retire from his professorship with the understanding that the university would provide him with facilities to set up a research institute. Instead, he was given a poky little office and no further support.[34]

But having his legs cut from under him as a researcher could not suppress Yudkin. In 1972 he published a scaremongering book on sugar's supposed dangers called *Pure; Sweet and Dangerous; The New Facts about the Sugar You Eat as a Cause of Heart Disease, Diabetes, and Other Killers* that sold very well in both Britain and America.[35] As a result, in 1974 embattled American egg producers brought him over for a media tour. There he repeatedly pointed out that not one study had ever demonstrated a cause-and-effect relationship between dietary cholesterol and heart attacks. The head of the AHA responded by saying that studies were in the works that would do this and reiterated its warnings against eating more than three eggs a week.[36] When Yudkin pointed out that many countries with elevated levels of heart disease also had high levels of sugar consumption, the AHA responded with a curt rejoinder saying that there was no experimental evidence proving that the two were correlated.[37] (Of course, this was also true of their diet/heart theory.)

Despite the AHA's clever attempt to tar him with the same brush he applied to them, Yudkin did manage to reinforce sucrophobia in America. It had originally received a major boost in 1970, when the Senate nutrition committee aroused fears that sweetened breakfast cereals were making children addicted to sugar, causing hyperactivity and deleterious health consequences for the rest of their lives. It then widened its net, hearing testimony on how the addiction began with sweetened baby foods, moved on into hard stuff in the Kellogg's boxes, and ended up creating a nation of sugar addicts. In his 1975 best-seller *Sugar Blues*, William Dufty invoked comparisons to heroin, calling sugar addiction "the white plague" and confessing that his first taste of it had led him down "the road to perdition"—stealing from his mother to buy candy to feed his addiction.[38]

Ultimately, though, the sucrophobes were no match for the lipophobes. While Yudkin, Dufty, and the rest ultimately had little impact on the American sweet tooth, the lipophobes' had an enormous effect on its fat tooth.[39] From 1956 to 1976, per capita butter consumption fell by over half and egg consumption dropped by over a quarter. Consumption of margarine doubled from 1950 to 1972 and that of vegetable oil rose by over 50 percent in the ten years from 1966 to 1976.[40] Given the slow pace at which consumption of core foods in a national diet normally changes, these are very impressive statistics.

In 1976 the lipophobes capped their apparent victory by bringing the U.S. government on their side. That year the Senate committee on nutrition began hearings on "Diet Related to Killer Diseases." The committee chairman, Democratic senator George McGovern, was already a convert to the diet-heart theory, as was its senior Republican member, Charles Percy. Percy had been much impressed by the findings of Alexander Leaf, a Harvard gerontologist who had returned from the Hunza Valley in 1973 and attributed the Hunzakuts' extraordinary longevity to, among other things, the paucity of animal fats and dairy products in their diets. The intrepid senator had then trekked there himself. On his return, he published an article in *Parade* magazine, titled "You Live to Be 100 in Hunza," confirming that the secret to Hunza longevity was plenty of exercise and the meager amount of animal fat in their diet.[41]

The committee's bias was revealed at the very outset, when it elicited testimony saying that an incredible 98.9 percent of the world's nutrition researchers believed that there was a connection between blood cholesterol levels and heart disease.[42] After a mere two days of hearings, Nick Mottern, an ex-labor reporter with no scientific training who lionized Keys, was assigned to write up the committee's report. The result, published early the next year as *Dietary Goals for Americans*, enshrined the diet-heart dogma into national nutrition policy. It called for Americans to increase their consumption of carbohydrates and to decrease their consumption of fats by 25 percent. Saturated fats were to be cut even more, by over one-third, mainly by cutting back on red meat.[43]

Livestock producers, alarmed that this would require a 70 percent reduction in meat consumption, managed to have the final report replace the call to severely limit red meat with one to "choose meats, poultry, and fish which will reduce saturated fat intake."[44] But the report was already preaching to the converted. That year an amazing three out of four Americans told pollsters that they were concerned about the amount of cholesterol in their diets.[45] And worry they might. By then, the AHA and others were telling them to limit their average daily cholesterol intake to less than 300 milligrams—not much more than the 250 milligrams in one large egg yolk.[46]

The beef producers' success in having the dietary guidelines altered ended up doing them little good. The year 1976 turned out to be the historic high point for American beef consumption. It declined by about

30 percent over the next fifteen years, before leveling off to its current level.[47]

"Cholesterol Is Proved Deadly"

In the late 1970s, lipophobes were hardly fazed by evidence that serum cholesterol was not uniformly deadly: that the high-density lipoproteins (HDL) that transport it were likely beneficial and that it was the low-density ones (LDL) ones that were the culprits. Instead, they continued to advocate the same low-fat diets on the grounds that reducing total cholesterol would reduce LDL. Doctors continued testing total cholesterol levels, and estimates of the levels at which these were said to put one "at risk" continued to be lowered.[48]

If anything, villainizing cholesterol picked up steam. In 1984 the National Institutes of Health announced that a long trial indicated that reducing cholesterol in the blood significantly reduced the risk of heart attacks. Although it was a drug, and not diet, that had reduced the test subjects' cholesterol, the agency now called on all Americans over the age of two to eat less fat. President Reagan promptly announced that he was giving up sausages and would henceforth drink only skim milk.[49] *Time* magazine ran a cover story, entitled "Cholesterol, and Now the Bad News," which began with the blunt statement, "Cholesterol is proved deadly." It said that the results of "the largest research project in medical history" were clear: "Heart disease is directly linked to levels of cholesterol in the blood" and "lowering cholesterol levels markedly reduces the incidence of fatal heart attacks."[50]

The next year the National Institutes of Health, supported by the AHA, mounted a National Cholesterol Education Program to persuade Americans to drastically cut back on cholesterol, "a major cause of coronary heart disease." It sent a large "Physicians' Kit" to every doctor in America, telling them to screen all their patients for cholesterol and to advise the vast majority of them to avoid saturated fat and replace butter with margarine.[51] In July 1988 the Surgeon General of the United States, C. Everett Koop, topped this by issuing a 700-page report that, in the words of yet another *Time* magazine cover story, "exhorts Americans to cut out the fat." It said that the evidence that fat was responsible for two-thirds of all the deaths in the United States was "even more impressive than that for tobacco and health in 1964." The president of the AHA said that if everyone went along with its recommendations, atherosclerosis would be "conquered" by 2000.[52] The AHA again urged mass screening of cholesterol levels, leading *Cardiovascular News* to tell physicians to expect "a flood of patients seeking treatment," which meant putting them on low-fat diets.[53]

Although it was now calculated that even a drastic reduction of dietary cholesterol would reduce the blood cholesterol levels of only about half of Americans, and then only by about 10 percent, the AHA and its allies continued to recommend low-fat diets for everyone. In 1987 and 1988, the AHA and the National Academy of Sciences, the Surgeon General of the United States, the National Heart, Lung and Blood Institutes, the National Cancer Institute, the U.S. Department of Agriculture, the Centers for Disease Control, the AMA, and the American Dietetic Association all "urged Americans *from age two upward* to go on restricted [low-fat] diets in the hope of preventing CHD [coronary heart disease.]" The AHA and Surgeon General urged food processors to help out by producing more low-fat foods.[54] In December 1988 *Time* magazine concluded another long cover story on HDL and LDL by quoting the president of the AHA as saying that "more than half the adult population" could reduce their chances of getting heart disease by lowering their LDL with low-fat diets. This included eating more low-fat processed foods, whose impact on heart disease would later be questioned.[55]

Stoking Women's Worst Fears

A significant obstacle to arousing the entire nation about the dangers of cholesterol was that heart attacks seemed to occur mainly in men. Indeed, in 1977 suspicious feminists suggested that it was no coincidence that the "killer disease" that most alarmed the all-male Senate nutrition committee was heart disease.[56] Yet men are not usually as compliant with diets as women, and they tend to be less guilt-ridden than women about what they eat. On the other hand, large numbers of middle-class women were already going on low-fat diets and buying low-fat foods to lose weight. It was quite easy, then, to add health concerns to their reasons for not eating fatty foods. In the 1980s the American Cancer Society, the AHA's erstwhile competitor for contributions, deftly exploited this by issuing warnings that breast cancer was linked to saturated fats.

Typically, this idea originated with researchers who found that countries with high rates of breast cancer had high levels of saturated fat in their diets.[57] The National Research Council then issued a report called "Diet, Nutrition and Cancer" that said that eating less fat would "likely reduce the risk of cancer."[58] The American Cancer Society promptly came out with an "anticancer" diet that called for eating less saturated fat. Its spokesman admitted that there was no substantive evidence to support the diet but assured consumers that it couldn't be harmful and would at least help combat heart disease.[59] In late 1986 a study of close to 90,000 nurses found absolutely no correlation between fat consumption and breast cancer, yet the Cancer Society continued its anti-fat crusade, adding more nails to saturated fat's coffin.[60]

The Mediterranean Diet to the Rescue

By 1990 the repeated failure of studies trying to prove that saturated fat caused heart disease was beginning to cause defections from the lipophobe side. In 1989 Harvard's Frederick Stare, who had enthusiastically supported Keys, reversed himself and co-authored a book denouncing the "cholesterol scare."[61] More challenges to lipophobia followed, culminating in 2006 with a massive study done as part of the NIH's Women's Health Initiative. It indicated that low-fat diets had no effect on rates of either cancer or cardiovascular disease among women.[62] Yet the lipophobes proved to be remarkably adept at bobbing, weaving, and altering their message in the face of the challenges.

The revival of Ancel Keys's Mediterranean Diet provided them with a handy way of doing this. In 1975 Keys and his wife revised their 1959 recipe book, entitling it *How to Eat Well and Stay Well the Mediterranean Way.* It aroused little interest until 1980, when the *Seven Countries* study finally came out. This provided conclusive proof, said Keys and his followers, of the correctness of his initial observations about the Mediterranean diet and heart disease.[63] Then, in the early 1990s came a surge in press coverage of the healthful qualities of the Mediterranean Diet. Much of this was due to the largesse of the International Olive Oil Council, which organized lavish conferences for food writers and other opinion leaders where experts credited the Mediterranean Diet with preventing heart attacks and cancer.[64] Although Ancel and Margaret Keys had hardly mentioned olive oil in their book, copious consumption of olive oil was now called the Mediterranean Diet's key protective element. It was now defined as "the diets of the early 1960s in Greece, southern Italy and other Mediterranean regions in which olive oil was the principle source of fat." Some faculty members at Harvard's School of Public Health were recruited to construct a new food pyramid, to replace the official U.S. government one. In the new Mediterranean pyramid, grains were at the bottom, as foods to be eaten most, and fruits and vegetables constituted the tier above them. Then olive oil was inserted, occupying a large tier all by itself. This was followed by diminishing tiers of other, less important, foods. Much to the delight of the International Olive Oil Council, olive oil sales soared, as did those of pasta.[65]

In late 1991 the American advocates of the Mediterranean Diet seemed to face a serious challenge to their demonization of saturated fat from across the sea. That November the influential television program *60 Minutes* opened with a shot of one of its hosts, Morley Safer, sitting in a restaurant in Lyons, France, facing platters of pâtés and sausages. He then posed the question of the "French Paradox." Why, he asked, did the French eat as much if not more saturated fat than Americans, and yet have a mortality rate from heart disease that was less than one-half of that of Americans? Raising a glass of red wine, Safer said, "The answer may well lie in this inviting glass." He then interviewed a number of French researchers who confirmed that, yes, this did indeed seem to be the case.[66]

Although red wine sales immediately skyrocketed, to Michel de Lorgeril, co-author of the "French Paradox" study, its most important aspect was that it scored a direct hit on Keys's dietary cholesterol/heart disease hypothesis.[67] Yet this was ignored in America, as Keys's followers responded by simply adding a recommendation that people who were already drinkers could "drink wine in moderation."

De Lorgeril then attacked from a different direction. He conducted a four-year-long experiment in which one group of heart attack survivors were put on a Mediterranean Diet that was high in fruits, vegetables, legumes, fish, and poultry, along with cholesterol-rich red meat and dairy products. Another group followed their doctors' recommendations for what the AHA called a "prudent diet," which mainly involved cutting back on saturated fat. After four years, the men on the "prudent diets" had a more than 50 percent higher rate of fatal heart attacks and 70 percent more cardiac events than those on the higher-cholesterol French-style Mediterranean Diet. Most significant to de Lorgeril was that the blood of both groups had almost identical levels of total cholesterol, LDL, and HDL.[68]

Yet when the study's final results were published in 1999 (in the AHA journal, no less) Americans steadfastly ignored the obvious implication that cholesterol did not cause heart disease. Instead, it was reported as proof of the efficacy of the American-style Mediterranean Diet, a key component of which remained an abhorrence of saturated fat. The study, said Jane Brody, the *New York Times* health columnist, was yet more substantiation for the wisdom that Ancel Keys had imparted "over a generation ago."[69]

By then, Keys, now long retired, was still basking in plaudits.[70] When he died in 2004, at age one hundred, he was hailed as the man who discovered the dangers of dietary fat and the lifesaving qualities of the Mediterranean Diet. Four years later the Greek, Italian, Spanish, and Moroccan governments persuaded UNESCO, the UN agency in charge of cultural preservation, to place the Mediterranean Diet on the World Heritage List. Some eyebrows were raised about the Greeks' leading role in this, as they were consistently ranked as the fattest of Europeans. Indeed, even Keys's paragons of Mediterranean eating, the residents of Crete, were discovered to have one of the highest levels of childhood and adolescent obesity in Europe.[71] This did not deter the Cretan tourist board from trying to have the Mediterranean Diet renamed the Cretan Diet.[72]

Big Checks for "Heart Checks"

Meanwhile, the American Heart Association continued to find new ways to prosper from lipophobia. In 1988 it deleted the provision in its charter prohibiting product endorsements and began offering, for a fee, to endorse any food products that met its guidelines for fat, cholesterol, and sodium. This allowed foods to carry a special AHA "Heart Guide" seal on their labels, with the logo "American Heart Association Tested and Approved." (Some eyebrows were raised when Rax restaurants began using the seal on its Big Rax roast beef sandwiches, whose thirty grams of fat were half of the AHA's entire daily allowance for a woman.) However, after the FDA objected that this implied that these foods were health foods, the AHA dropped the program until the FDA came out with new rules for food labels.

Once the new rules came out, the AHA remounted the campaign. This time it sold the right to use a "Heart Check" symbol and say "Meets American Heart Association food criteria for saturated fat, cholesterol and whole grains for healthy people over age 2." For this, it charged fees ranging from the $2,500 it cost Kellogg's for each of the more than fifty of its products which qualified (including such nutritional dazzlers as Fruity Marshmallow Krispies) to the $200,000 that Florida citrus fruit producers paid for exclusive rights to the symbol, cutting out their competitors in California. The Florida producers now ran ads saying, "Fight Heart Disease. Drink Florida Grapefruit Juice." These pictured a jug of juice, the AHA Heart Check, the words "Certified Heart Healthy," and a heart enclosing the phrases "Cholesterol Free" and "Fat Free." The AHA's old bête noire, the Cattlemen's Association, not only bought the right to put the Heart Check seal on beef; it was also received a special "Champions of Heart" award for unspecified "contributions that . . . allowed AHA to further its fight against heart disease and stroke." In 1992–93 ConAgra, the hydra-headed giant involved in practically every stage of food production, gave $3.5 million to the AHA, ostensibly to make a television program on nutrition.[73]

Yet by the end of the century, the AHA's calls to reduce heart disease through diet were sounding rather threadbare. It seemed that only in extreme cases, such as those terrified men who were able to stick with the incredibly restrictive Dean Ornish diet, could people stay on a diet low enough in saturated fat long enough for it to have much of an effect on their LDL/HDL levels. For everyone else, there was still no evidence that low-fat diets prevented heart disease. In 1996 the American College of Physicians came out against the AHA program of screening all people over twenty for high cholesterol. It said that it resulted in young people being put on low-fat diets that rarely reduced cholesterol. They were then told to take medications, whose long-term effects were unknown, for the rest of their lives. Others began pointing out that the AHA campaign to have people adopt low-fat, high-carbohydrate diets led to increased consumption of calorie-dense foods that contributed to obesity and diabetes, both of which were risk factors for cardiovascular disease.[74]

However, in 2000 another panic gave lipophobia yet another boost. This time it was about trans fats, which were in the hydrogenated oils used in making everything from French fries to Doritos to granola bars. Not only did they raise levels of "bad" LDL in the blood; they also lowered "good" HDL ones. New York City banned trans fats from restaurants, school boards across the nation banished them from cafeterias, and processors began furiously trying to replace them.[75]

One would think that the trans fat scare might prompt lipophobes to eat a bit of humble pie. After all, years before, when they were first raised, Keys had dismissed these fears as groundless. Then, in the 1970s, the AHA, the Center for Science in the Public Interest, and other agencies had urged processors to use trans fats to replace supposedly deadly saturated fats.[76] Moreover, the most common conveyor of trans fats to the bloodstream turned out to be margarine, which they all had recommended as a heart-healthy alternative to deadly butter.[77] Yet nary a *mea culpa* was heard.

Florida citrus growers were among the many food producers who tried to profit from lipophobia—fear of dietary fat—by paying the American Heart Association to use its "Heart Check" symbol to support questionable claims that their fat-free or low-fat products helped prevent heart disease.

However, at least all the talk about "good" and "bad" fats finally forced lipophobes to abandon their calls for reducing total consumption of dietary fat. Instead, the AHA began recommending that people substitute unsaturated fats, such as olive oil, for the saturated and trans fat ones in their diets. Similarly, in 2000 the government's revised *Dietary Guidelines* replaced the previous advice to choose a diet that was "low in fat, saturated fat and cholesterol" with a diet that was "low in saturated fat and cholesterol and moderate in total fat." However, the baleful effects of lipophobia could not be eradicated so easily. In 2001 a number of repentant lipophobes expressed regret that the long campaign they had waged against total fat had "made the belief that fat is bad so strong and widespread" that it would take a herculean effort to undo it.[78]

The AHA and Big Pharma

By then, food producers were facing new competition for the "food-fear" dollar from their old rivals, the pharmaceutical companies. In 1987 the FDA approved the use of a class of drugs called statins that markedly lowered the amount of cholesterol in the blood. By the mid-1990s millions of Americans were taking them, and they seemed to be providing some protection against heart disease.[79] By 2004 the AHA was acknowledging their effectiveness.[80] Perhaps coincidentally, that same year Merck and Schering-Plough—a pharmaceutical joint venture that produced Vytorin, a cholesterol-reducing drug that combined the statin Zocor with another drug—began contributing $2 million a year to the AHA. It also paid $350,000 a year to sponsor the "cholesterol page" on the AHA website, which featured a direct link to the Vytorin site. In return, the AHA helped Merck and Schering-Plough with their $150 million-a-year marketing campaign for the drug. All of this seemed to pay off handsomely. Soon, over 4 million Americans were taking Vytorin or the other non-statin, Zetia, bringing $5 billion a year into the two drug companies' coffers.[81]

In January 2008 Schering-Plough's investment in the AHA seemed to pay off in another way. A disturbing study showed that Vytorin was no better at reducing arterial plaque than Zocor alone and might actually speed up the process. Yet Vytorin cost $100 for a thirty-day supply while Zocor only cost $6. When the authoritative *New England Journal of Medicine* supported the study's recommendation that patients use Vytorin only as a last resort, the AHA quickly rallied to Schering-Plough's defense by condemning the study as too limited and warning people not to stop taking Vytorin without consulting their doctors. This advice looked quite ill-advised some months later, when a study showed that not only did Vytorin not reduce "bad" LDL cholesterol, but in some cases may have increased it.[82]

Endorsing drugs hardly deterred the AHA from continuing to tell Americans to avoid heart disease by changing their diets. In 2000 it had introduced a new "heart-healthy" diet that, like its older ones, recommended eating plenty of no-fat and low-fat foods to reduce cholesterol levels. It also continued to endorse "heart-healthy" low-fat processed foods to help them do so, collecting over $15 million for these endorsements in 2007.[83] Yet by then, even lipophobic scientists were saying that it was cholesterol in the blood, not in the diet, "that counted."[84]

Then, in late 2008 came an apparently crushing scientific blow. A new study seemed to explain something that the diet-heart theory could not: namely that half of heart attacks and strokes occur in people with low or normal LDL cholesterol levels. The new theory claimed that the main culprit in heart disease was not fat, but inflammation. Statins were effective, it said, because they reduced levels of a protein, called high-sensitivity C-reactive protein (CRP), that contributes to inflammation in the body. The crucial risk factor for heart disease was therefore not cholesterol, but elevated CRP, which has nothing to do with fat in the diet.[85] In July 2009 another study tried to alter this theory by reducing CRP to the role of an indicator, not a cause, of heart disease. Inflammation remained a villain, but whether it was cause or effect was unknown. Cholesterol's role in all of this was still unclear, but it seemed highly unlikely that Keys's diet-heart theory would ever be resurrected.[86]

However, fear of saturated fat was already too ingrained in the popular mind to be shaken. In February 2010 the press reported on a meta-analysis of twenty-one lengthy studies, comprising 347,747 subjects, that concluded that there was no association between saturated fat consumption and the risk of heart disease.[87] Two days later, in reporting on ex-president Bill Clinton's hospitalization because of a clogged artery, the Associated Press pointed out that he was "a legend as an unhealthy eater," with a weakness for saturated fat–laden hamburgers and steaks.[88]

Unintended Consequences

Lipophobia shared a number of characteristics with other modern food scares. Like germophobia and the milk scare, it originated in fear of an epidemic. As with the thiamine scare, a lack of solid evidence did not deter its proponents from calling for wholesale changes in the American diet. Powerful commercial interests profited handsomely from the scare, just as they did from vitamania and natural foods. Like vitamania, it exemplified the dangers of recommending dietary changes for an entire nation, rather than the much smaller proportion of people who may be seriously at risk.

However, vitamania just led to the needless expenditure of billions of dollars on mainly harmless vitamins. In contrast, lipophobia may well have done considerable harm. Indeed, had John Yudkin, who died in 1995 at age eighty-four, lived, like Keys, to be one hundred, he may well have felt that he was having the last laugh. First, proponents of high-protein diets revived his claims that diets high in refined carbohydrates contributed to obesity and heart disease. They were especially critical of the low-fat processed foods promoted by the American Heart Association. They pointed out that the fats in them were replaced by high-calorie sweeteners and carbohydrates, which were now said to contribute to metabolic syndrome, a major risk factor for diabetes, heart disease, and other serious illnesses.[89] In 2009 the AHA itself issued a special report calling on most Americans to drastically cut their consumption of added sugars. Why? It said that the excessive amount of sugar in their diets contributed to obesity, high blood pressure, and other health problems that increased the risk of heart attack and stroke.[90]

I cannot end without mentioning that the most obvious risk factor for heart disease, as well as many other "killer diseases," seems to be poverty. Despite the earlier perceptions that heart disease was the wealthy man's disease, the fact is that death rates from heart disease—that is, the chances of dying from it in a given year—increase with lower socioeconomic status. This is the case for women (whose heart disease rates catch up to men's after about age sixty) as well as men. There is no consensus about why this is so, but there is considerable irony in the fact that there is considerable support for one explanation, put forward by the British epidemiologist Michael Marmot, that blames stress. In a reversal of the 1950s idea that successful businessmen experience the most stress, he says that the lower one's status, the less control one has over many of the most important aspects of one's life. The result is a much more stressful daily life for the poor than for the well-off, with devastating effects on their health.[91]

Whether or not Marmot is right about the causes, the fact that low socioeconomic status is the greatest risk factor for heart disease would seem to indicate that the best way to reduce deaths from it would be to reduce poverty. But of course there are few experts who advocate this solution. To do so would run counter to the whole tenor of the times, in which Americans search for individual, rather than collective, solutions to their problems, and seek to blame people's afflictions on their lifestyle choices, rather than on their social circumstances or the luck of the draw.

Coda

Surely the most striking thing about this story is how markedly ideas about food's healthfulness have changed over the years. Chemical preservatives went from being triumphs of modern science to poisons. Whole milk has swung back and forth like a pendulum. Yogurt experienced boom, bust, and revival. Processed foods went from bringing healthy variety to the table to being devoid of nutrients. Prime rib of beef was transformed from the pride of the American table into a one-way ticket to the cardiac ward. Margarine went from "heart-healthy" to artery-clogging, and so on. As of this writing, we are told that salt, historically regarded as absolutely essential to human existence, is swinging the grim reaper's scythe.

These fluctuations have largely been the result of shifts in scientific opinion that various sources—public, professional, and commercial—publicized among the middle classes. As the years went by, the pace of change in experts' advice accelerated and the number of agencies disseminating their ideas multiplied. In 1993 I quoted Claude Fischler as calling the resulting din of often-contradictory messages a "nutritional cacophony." He suggested that this might result in a kind of "gastro-anomie," a condition in which people have no sense of dietary norms and rules. Today, however, nothing like that appears to be on the horizon.[1] Yes, the shifting sands of nutritional advice have aroused frequent expressions of skepticism, if not cynicism. The wry comment that "for every Ph.D. there seems to be an equal and opposite Ph.D." still packs a punch. However, this skepticism seems to last only until the next great scare. Then, once again, a confluence of scientific expertise, puritanical instincts, and, yes, pecuniary interests arises to play on the anxiety that so many Americans feel as they contemplate what they are about to eat.

It all seems to represent rather sad confirmation of what I suggested in the introduction to this book: that the forces of modern science, industrialization, and globalization have combined with the "omnivore's dilemma" and Americans' individualistic ethos to turn fear of food into something akin to a permanent condition of middle-class life. Repeatedly, the fearmongers, well-oiled by money and scientific prestige, have steamrollered over those few who have argued that there are no certain answers to questions about food and health—that food and humans are just too variable for that. Hopefully, this description of the origins and nature of this process will help the reader to avoid being flattened. Perhaps it will also come to the attention of some of a new generation of experts, whose knowledge of the history of science and medicine often consists of triumphalist tales of how brilliant researchers made discovery after discovery on the road toward unlocking the mysteries of human life. Some familiarity with the real history of their fields might give them a much-needed dose of caution about telling people how to eat. And who knows, the book might also spur some of the journalists who breathlessly report their findings to approach them with a more critical eye.

During the course of writing this book, I have often been asked what lessons I personally have drawn from it. Well, for me, the history of expert advice on diet and health inevitably brings to mind the old saying "This too shall pass." The massive reversals in expert opinions described in this book provide more than enough support for this skepticism. Indeed, the hubris of many of the experts confidently telling us how to eat has often been well-nigh extraordinary. In 1921, for example, the consensus among the nation's nutritional scientists was that they knew 90 percent of what there was to know about food and health. Yet just a few years earlier, before the discovery of vitamins, they had routinely condemned the poor for wasting their money on fresh fruits and vegetables, which were said to be composed of little more than water, with only minimal amounts of the protein, fats, and carbohydrates that were essential to life.[2] Nowadays, although the experts do admit that much remains to be discovered, they still exude unwarranted confidence in their pronouncements, despite the fact that the small print inevitably includes crucial hedges, such as the omnipresent word "might" and phrases such as "may be related to." So, when apprised of new nutritional discoveries, I tell myself to wait, wait, and then wait some more, before even thinking of making important changes in how I eat.

Second, I always remind myself that the experts have an unfortunate tendency to come out with one-size-fits-all recommendations. This book contains a number of examples of their telling all Americans to avoid eating things that, even by their calculations, only a minority of them should be concerned about. They seem to assume that finely parsed recommendations—ones that apply only to certain categories of people deemed to be "at risk"—are not easily understood by the general public. Nor does advice with qualifications make for good news coverage or catchy advertising copy.

I am also cautious about following advice to deny myself foods I like. This book shows how countless people were subjected to unnecessary deprivation and/or feelings of guilt because of scientific certainties that were later found to be untenable. I often think of my late father-in-law, who spent his last twenty-odd years grimly following doctors' orders to deny himself the beef, cheese, butter, and eggs he so dearly loved. As the two final chapters show, there was probably no need for him to have been deprived of so much pleasure.

At the same time, I have become more sensitive to how dubious ideas about diet and disease can encourage blaming the victims. I now feel pangs of guilt over my reaction to my sister's death from cancer in the late 1980s, when I could not help but think that she might have brought it on herself. Why? This was when the American Cancer Society was saying that there was a direct link between eating saturated fat and breast cancer, and there were few things my sister enjoyed more than a thick corned beef sandwich. The fact that this link was later shown to be nonexistent made me resolve never to go down that path again.

How, then, can one avoid being swept up by these food fears? I would say that the first tactic is to look at those propounding these fears and ask, "What's in it for them?" As this book shows, "them" of course includes the usual suspects: food companies trying to promote and profit from food fears. However, it also includes thousands of other people with career interests in scaring us. This means not just scientists hoping to keep the research grants flowing by discovering connections between diet and health; it also involves well-meaning people working for public and nonprofit agencies who try to prove their usefulness by warning about dangerous eating habits. As my father used to remind me, even those with lofty motives "still have to make a living."

I also try to bear in mind how often moralism, rather than science, underlies food fears. Calls for self-denial inevitably tap into the Puritan streak that still runs deep in American culture. Left-wing exposures of the dangers in the food supply often emerge from a worldview that sees the nefarious machinations of big business at the root of all that is wrong with America. Right-wingers tend to ignore the role of socio-economic factors in health outcomes and seem to gain some dark satisfaction from blaming ill health on morally weak individuals' poor food choices. All such attempts to inject morality into food choices amount to rejections of what may well really explain the "French Paradox": that the kind of pleasure that the French get from eating, and from sharing the experience with others, can contribute to good health.

What, then, to do in the face of all these calls to approach food with fear and dread? What should we eat and what should we avoid? In this regard, the kind of thing that Michael Pollan has been saying strikes me as eminently sensible: Eat a wide variety of foods, don't eat too much, and eat relatively more fruits and vegetables. This is a variation on the older advice that it's all right to eat everything "in moderation." This is what the cookbook author Julia Child called for toward the end of her life, when she was often condemned for advocating the liberal use of the butter, cream, and other pleasurable foods. Of course, her idea of moderation ran completely counter to that of most of her critics, but in this, as in many other respects, I'll side with Julia.

Abbreviations for Frequently Cited Sources

AJCN	*American Journal of Clinical Nutrition*
AJN	*American Journal of Nursing*
AJPH	*American Journal of Public Health*
BDE	*Brooklyn Daily Eagle*
G&M	*Globe and Mail* (Toronto)
GH	*Good Housekeeping*
JADA	Journal of the *American Dietetic Association*
JAMA	*Journal of the American Medical Association*
JN	*Journal of Nutrition*
NEJM	*New England Journal of Medicine*
NT	*Nutrition Today*
NYT	*New York Times*
NYTM	*New York Times Magazine*
OGF	*Organic Gardening and Farming*
SNL	*Science News Letter*
ToL	*Times of London*
USDA	United States Department of Agriculture
WP	*Washington Post*

Notes

Preface

1. "The Wonders of Diet," *Fortune*, May 1936, 86.
2. Harvey Levenstein, *Seductive Journey: American Tourists in France from Jefferson to the Jazz Age* (Chicago: University of Chicago Press, 1998); *We'll Always Have Paris: American Tourists in France since 1930* (Chicago: University of Chicago Press, 2004).
3. Paul Rozin, Claude Fischler, Sumio Imada, Alison Sarubin, and Amy Wrzesniewski, "Attitudes to Food and the Role of Food in Life in the U.S.A., Japan, Flemish Belgium and France: Possible Implications for the Diet-Health Debate," *Appetite* 33 (1999): 163–80; Claude Fischler and Estelle Masson, *Manger: Français, Européens et Américains face à l'alimentation* (Paris: Odile Jacob, 2007), 37. Perhaps because Europe lagged behind the United States in industrializing its food supply, it now seems to be playing catch-up when it comes to worries over the consequences of the chemicals, preservatives, and processes involved. Recent polls indicate that these concerns are now increasing there as well. Claude Fischler, personal communication with the author, October 10, 2009. Canadians, whose foodways have been shaped by many of the same forces as Americans', have followed even more closely. This is why, although I am Canadian, I allowed my American publisher to persuade me to include "we" in the book's title.
4. Harvey Levenstein, *Paradox of Plenty: A Social History of Eating in Modern America* (New York: Oxford University Press, 1988; rev. ed., Berkeley: University of California Press, 2003), 255. The first volume is *Revolution at the Table: The Transformation of the American Diet* (New York: Oxford University Press, 1993; 2nd ed., Berkeley: University of California Press, 2003).

Introduction

1. Paul Rozin, "The Selection of Foods by Rats, Humans, and Other Animals," in *Advances in the Study of Behavior*, vol. 6, ed. J. S. Rosenblatt, R. A. Hinde, E. Shaw, and C. Beer (New York: Academic Press, 1976), 21–76. Fischler then wrote of it as the "omnivore's paradox." Claude Fischler, *L'Homnivore: Le goût, la cuisine et le corps* (Paris: Odile Jacob, 1990). Michael Pollan credits Rozin for the term. Michael Pollan, *The Omnivore's Dilemma: A Natural History of Four Meals* (New York: Penguin, 2006), 3.

2. Madeleine Ferrières has chronicled the way in which the market economy contributed to periodic food fears with regard to meat in early modern and nineteenth-century France. Madeleine Ferrières, *Sacred Cow, Mad Cow: A History of Food Fears*, trans. Jody Gladding (New York: Columbia University Press, 2006).

Chapter One

1. The theory was also the result of the work of a number of scientists, as well as physicians such as the German Robert Koch. See Nancy Tomes, *The Gospel of Germs: Men, Women, and Microbes in American Life* (Cambridge, MA: Harvard University Press, 1998), 23–47. Pasteur initially gained fame in the 1870s when he helped save the French brewing and wine industries by inventing ways to deal with the microbes that were spoiling their products. His research on sick silkworms then led him to conclude that each disease is caused by a specific microbe, or bacterium—an insight that had immense implications for human ailments.
2. *NYT*, February 3, 1895.
3. The zoologist attributed the slothfulness of poor whites in the South to a laziness-causing germ that thrived in their unsanitary food and environment. *BDE*, December 5, 1902. Actually, he was not all that far off the mark, for the lassitude was a symptom of pellagra, a vitamin-deficiency disease caused by the poor diet based on cornmeal, and of hookworm, which flourished in the region's primitive sanitary facilities.
4. *WP*, August 16, 1908. This involved a complete turnabout from 1903, when Wiley said that incidence of male baldness was increasing because men were becoming more intelligent. As their brains grew larger, they took nutrients from their hair, which then fell out. Women, on the other hand, had long hair because "woman is still a savage.... Her brain hasn't the capacity of man's." Women's savagery, he said, was proven by their fondness for "gaudy colors." *WP*, September 17, 1903.
5. Phyllis A. Richmond, "American Attitudes toward the Germ Theory of Disease," *Journal of the History of Medicine and Allied Sciences* 9 (1954): 428–54. By 1890, though, most doctors were coming around to it. H. W. Conn, "The Germ Theory as a Subject of Education," *Science* 11, no. 257 (January 6, 1888): 5–6.
6. *NYT*, May 11, 1884; February 17, 1895; June 21, 1914.
7. *NYT*, April 23, 1893.
8. *BDE*, April 14, 1893; January 31, 1897. By 1910 germophobia was being used to sell everything from antiseptic wall paint to household cleansers and toothpaste. Nancy Tomes, "The Making of a Germ Panic, Then and Now," *AJPH* 90 (February 2000): 193.
9. *NYT*, April 22, 1894.
10. *NYT*, February 2, 1913.
11. *NYT*, October 7, 1902. The first attempt to measure them failed because the gelatin plates used to "catch" the germs were so overloaded that it was impossible to distinguish between the various kinds. Nevertheless, the street cleaning commissioner concluded that there was "a positive danger in every peddler's cart in the city." *BDE*, July 24, 1902.
12. Joel Tarr and Clay McShane, "The Centrality of the Horse to the Nineteenth-Century American City," in *The Making of Urban America*, ed. Raymond Mohl (Wilmington, DE: Scholarly Resources, 1997), 105–30.

13. Judith Walzer Leavitt, *Typhoid Mary, Captive to the Public's Health* (Boston: Beacon Press, 1996); Anthony Bourdain, *Typhoid Mary: An Urban Historical* (New York: Bloomsbury, 2001).
14. *NYT*, November 28, 1909; May 19, 1912.
15. It was recited by Nucky Thompson, the leading character in the TV series *Boardwalk Empire*, in the episode broadcast on November 1, 2010. According to various web sources, the song was cowritten around 1913 by Roy Atwell, a vaudeville performer who later did the voices for cartoon characters in Disney films. It was recorded in 1915 by a singer named Bill Murray and in 1946 by the radio star Phil Harris.
16. Helen S. Gray, "Germophobia," *Forum* 52 (October 1914): 585.
17. *NYT*, February 2, 1913; July 26, 1916.
18. *BDE*, January 3, 1901; "Major Walter Reed," Walter Reed Army Medical Center website, http://www.wramc.amedd.army.mil/visitors/visitcenter/history/pages/biography.aspx. Warnings about flies' ability to spread disease were sounded as early as 1883, when a Dr. Thomas Taylor warned that their propensity to flit between manure pile and milk pail spread any number of illnesses. Tomes, *Gospel of Germs*, 99.
19. *NYT*, December 24, 1905. A not surprising figure, even had it been true, since the United States lost only 379 men in combat and more than 5,000 to disease. By 1917 the Boer War, with many more casualties, had been added to the indictment. A *New York Times* editorial calling for support of the "swat the fly" movement claimed that "flies killed more men than bullets in the Boer and Spanish-American wars." *NYT*, July 29, 1917.
20. *NYT*, March 19, 1908.
21. Proof of this danger was not exactly compelling. It was that "wherever in any part of the world tuberculosis is present there also is the fly found as a pest." *NYT*, June 18, 1905.
22. *NYT*, December 24, 1905; July 26, August 10, 1908; June 28, August 21, 1910. Later a New York City doctor warned that bedbugs transmitted tuberculosis. *NYT*, December 12, 1913.
23. *NYT*, December 24, 1905.
24. Following hard on this success, Crumbine zeroed in on germs spread by water drunk in public places from common drinking cups and pushed successfully for them to be replaced by his paper "Health Cup," later called a Dixie Cup. He ended his days as a consultant to the paper cup industry. Robert Lewis Taylor, "Swat the Fly," *New Yorker*, July 17, 1948, 31–39; July 24, 1948, 29–34; *Newsweek*, October 11, 1948; Alan Greiner, "Pushing the Frontier of Public Health," Kansas University School of Public Health website, 2006, http://www.kpha.us/documents/crumbine_frontier.htm; Aileen Mallory, *Child Life*, October 2000.
25. *NYT*, March 19, 1908; April 18, 1909; March 24, 1913; Margaret Deland to C. F. Hodge, April 1, 1912, in *NYT*, May 19, 1912.
26. *NYT*, May 19, 1912; Tomes, *Gospel of Germs*, 145.
27. *NYT*, May 19, 1912.
28. *BDE*, June 29, 1902; *NYT*, June 21, 1914. The disinfectant industry had begun blossoming in the 1860s, riding the crest of the miasma theory. The rise of the germ theory necessitated little adjustment in its incredible panoply of products, which simply took to using Pasteur's name or, as in the case of Listerine, the name of the man who introduced asepsis to the hospital. Tomes, *Gospel of Germs*, 68–87.

29. Tomes, *Gospel of Germs*, 169.
30. "Admit It, Uneeda Biscuit," *Shockhoe Examiner*, August 7, 2009. http://theshockoe examiner.blogspot.com/2009/08/admit-it-uneeda-biscuit.html.
31. *NYT*, August 6, 1914.
32. Levenstein, *Revolution*, 40–41.
33. *NYT*, July 29, 1917.
34. *NYT*, February 12, 1911.
35. *NYT*, July 29, 1917.
36. *ToL*, October 1, 1920.
37. U.S. Department of Agriculture, *The House Fly and How to Eradicate It* (Washington, DC: USGPO, 1925).

Chapter Two

1. *NYT*, June 21, 1914.
2. See my *Revolution at the Table: The Transformation of the American Diet* (Berkeley: University of California Press, 2003), 121–36, for a detailed discussion of the controversy over artificial infant feeding.
3. A typhoid outbreak that struck the social luminary Cornelius Vanderbilt and a number of other wealthy New York City moguls in 1902 was immediately blamed on the city's water supply. *BDE*, December 24, 1902.
4. *BDE*, August 15, 1886; February 14, February 24, 1889.
5. *NYT*, September 22, 1895; August 27, 1899; July 27, 1910. The city health department had tried to persuade poor mothers to nurse their babies during the hot summer months when bacteria multiplied inordinately in milk but with little success. *NYT*, July 27, 1910.
6. *BDE*, October 9, 1900.
7. *NYT*, June 10, 1903.
8. *BDE*, October 9, 1900.
9. *NYT*, February 20, 1907. In fact, the bovine and human forms of the tuberculin are distinct. In Great Britain, the bacteria in impure milk were also blamed for appalling infant mortality rates, by causing diseases such as scarlet fever, tuberculosis, and severe infant diarrhea. P. J. Atkins, "White Poison?: The Social Consequences of Milk Consumption, 1850–1930," *Social History of Medicine* 5 (1992): 207–27.
10. U.S. Public Health Service, *Report No. 1 on the Origin and Prevalence of Typhoid Fever in the District of Columbia* (Washington, DC: USGPO, 1907); *NYT*, October 27, 1912.
11. Milton Rosenau, *The Milk Question* (Boston: Houghton Mifflin, 1912), 2; emphasis in original. Rosenau had been the lead author of the 1907 typhoid fever report, which was followed by another three volumes chronicling milk's dangers.
12. They also provided an equal number of cups of milk for consumption at the stations. *NYT*, July 30, 1893; *BDE*, June 11, 1899; October 2, 1902.
13. *NYT*, September 28, 1911; August 31, 1912; Clayton A. Coppin and Jack High, *The Politics of Purity: Harvey Washington Wiley and the Origins of Federal Food Policy* (Ann Arbor: University of Michigan Press, 1999), 145.

14. Irene Till, "Milk: The Politics of an Industry," in *Price and Price Policies*, Walton Hamilton et al. (New York: McGraw-Hill, 1938), 450–51.
15. Pediatricians, seeking to establish the credentials of their new specialty, developed a complex system of infant feeding called milk modification. This bit of hocus-pocus had them write individual prescriptions for babies, which were then taken to pharmacists or "milk stations," where the milk was diluted with water and cream and powders were added to bring about the supposedly optimum proportions of proteins, fats, carbohydrates, and casein for each baby. Poorer mothers who had to feed their children artificially often just diluted the cow's milk with water and perhaps added some sugar. Not only did this leave the children nutritionally deficient; the water was often full of bacteria that multiplied rapidly in the warm milk, causing many fatal cases of diarrhea. Levenstein, *Revolution*, 121–36.
16. *WP*, April 11, 1907.
17. *NYT*, February 20, 1907; May 19, August 31, 1912. It was thought, erroneously, that tubercular cows passed the tuberculosis bacillus into their milk.
18. Doctors opposed to pasteurized milk argued, with no proof, that it robbed milk of its "nutritive value." *NYT*, February 24, 1907; January 28, 1911.
19. *NYT*, December 8, 1910.
20. Leo Rettger, "Some of the Newer Conceptions of Milk in Its Relation to Health," *Scientific Monthly* 5 (January 1913): 64; USDA, Economic Research Service, "U.S. per capita food consumption, Dairy (individual), 1909–2004"; "Fluid milk and cream: per capita consumption, pounds, 1909–2004"; Judy Putnam and Jane Allshouse, "Trends in U.S. Per Capita Consumption of Dairy Products, 1909 to 2001," [USDA, ERS] *Amber Waves*, June 2003.
21. This is not say that pasteurization played more than a minor role in the steady decline in infant mortality during the first half of the twentieth century. In *Revolution at the Table*, I point out that the two have never been convincingly linked together. There was little difference in mortality rates between breast-fed and artificially fed infants. I argue that the decline in infant mortality was more likely linked to rising standards of living and better nutrition, especially among poor women. Levenstein, *Revolution*, 135–36.
22. *NYT*, January 13, April 4, 1920.
23. E. Melanie DuPuis, *Nature's Perfect Food: How Milk Became America's Drink* (New York: NYU Press, 2002), 106–9; *NYT*, June 5, 1921.
24. DuPuis, *Nature's Perfect Food*, 109–10.
25. Levenstein, *Revolution*, 154; DuPuis, *Nature's Perfect Food*, 110.
26. *NYT*, June 5, 1921.
27. DuPuis, *Nature's Perfect Food*; Levenstein, *Revolution*, 154–55; Elmer V. McCollum, *The Newer Knowledge of Nutrition*, 2nd ed. (New York: Macmillan, 1923), 398–415, 421. McCollum also contrasted the vigor of the descendants of the Loyalists who went to Canada after the American Revolution, who were "industrious, educated, and progressive," with those who went to the Bahamas, who were "improvident, lazy . . . indolent and unprogressive." This, he said, was because the Canadians consumed more dairy products. He did not mention that most of the lazy Loyalists who went to the Bahamas were Southern slave owners who took their slaves with them.

28. Levenstein, *Revolution*, 155; Richard O. Cummings, *The American and His Food: A History of Food Habits in the United States*, rev. ed. (Chicago: University of Chicago Press, 1941), 269; Levenstein, *Paradox*, 97; USDA, Economic Research Service, "Fluid milk and cream: Per capita consumption, pounds, 1909–2004."
29. Nancy Tomes, "The Making of a Germ Panic, Then and Now," *AJPH* 90 (February 2000): 191–94.
30. *Greensboro (NC) Record*, November 7, 2006; *Hamilton (Ont.) Spectator*, November 9, 2006.

Chapter Three

1. Indeed, one frightening calculation was that 1.28 billion dangerous new bacteria grew there each day. *San Francisco Call*, May 31, 1903.
2. Metchnikoff's description of how this affected Russian science was eerily prescient of what transpired in Germany after the Nazi takeover in the 1930s. He told a *New York Times* interviewer that as a result of the persecution, "the Russian universities had no serious professors, and some of the best men in Russia who could achieve much for their fatherland were lost to Russia. I am speaking of the Russian Jews. . . . Russia has lost many great talents by persecuting the Jews." *NYT*, August 1, 1909.
3. *NYT*, January 29, 1900.
4. *NYT*, December 21, 1907; August 1, September 19, 1909.
5. The theory that undigested elements in the body can cause disease can be traced back to the ancient Greek practitioners Galen and Hippocrates, and has persisted in Western medicine in some form or another since then. E. M. D. Ernst, "Colonic Irrigation and the Theory of Autointoxication: A Triumph of Ignorance over Science," *Journal of Clinical Gastroenterology* 24 (June 1997): 196–98.
6. *NYT*, June 20, December 12, 1903; Elie Metchnikoff, *The Nature of Man*, trans. Peter Mitchell (New York, 1903; reprint, New York: Arno Press, 1977), 60; R. B. Vaughan, "The Romantic Rationalist: A Study of Elie Metchnikoff," *Medical History* 9 (July 1965): 201–15; James Whorton, *Crusaders for Fitness: The History of American Health Reformers* (Princeton, NJ: Princeton University Press, 1982), 216–19; Alfred I. Tauber, "The Birth of Immunology: III. The Fate of Phagocytosis Theory," *Cellular Immunology* 139 (February 1992): 505–30; Thomas Söderqvist and Craig Stillwell, "Immunological Reformulations," *Science* 256 (May 15, 1992): 1050–52; Scott Poldolsky, "Cultural Divergence: Elie Metchnikoff's *Bacillus bulgaricus* Theory and His Underlying Concept of Health," *Bulletin of the History of Medicine* 72, no. 1 (1998): 1–27; James Whorton, *Inner Hygiene: Constipation and the Pursuit of Health in Modern Society* (Oxford: Oxford University Press, 2000), 173.
7. Elie Metchnikoff, "The Wilde Medal and Lecture," *British Medical Journal* 1 (1901): 1028, cited in Poldolsky, "Cultural Divergence," 4; *San Francisco Call*, May 31, 1903; Horace Fletcher, *The New Glutton or Epicure* (New York: Stokes, 1906), 177. The small intestine joins the stomach to the large intestine.
8. Whorton, *Inner Hygiene*, 173.
9. Elie Metchnikoff, "The Haunting Terror of All Human Life," *NYT*, February 27, 1910.
10. In 1909 Metchnikoff sent an assistant to study the bacteria in what was left of the in-

testines of forty of Lane's patients, who were reported to be in excellent health, and claimed to have found support for his theory in the low level of bacteria found there. *New York Sun*, July 11, 1909.

11. *San Francisco Call*, May 31, 1903; *NYT*, February 6, 1910. As an added benefit, this good bacillus would also protect against the bacteria that were thought to cause "arteriosclerosis, one of the main symptoms of premature old age." *NYT*, October 31, 1909.
12. Elie Metchnikoff, *The Nature of Man; Studies in Optimistic Philosophy* (New York: Putnam, 1903); Arthur E. McFarlane, "Prolonging the Prime of Life: Metchnikoff's Discoveries Show that Old Age May Be Postponed," *McClure's* 25 (September 1905): 541–51; William Dean Howells, "Easy Chair," *Harper's Monthly*, October 1904, 804.
13. *Washington Times*, October 20, 1904; McFarlane, "Prolonging," 549; *NYT*, August 16, 1908.
14. *NYT*, November 17, 1907.
15. Elie Metchnikoff, *The Prolongation of Life* (New York: Putnam, 1908); "Studies of Natural Death," *Harper's Monthly*, January 1907, 272–76; *NYT*, January 18, 1908.
16. *NYT*, January 10, 1910.
17. Loudon M. Douglas, *The Bacillus of Long Life* (New York: Putnam, 1911); *NYT*, January 21, 1912.
18. *NYT*, August 16, 1908, January 2, 1910; *ToL*, April 13, April 28, 1910; Douglas, *Bacillus of Long Life*.
19. The major cause of diabetes, they said, was "cold storage food." *NYT*, July 20, 1913.
20. Elie Metchnikoff, "Why Not Live Forever?" *The Cosmopolitan* 53 (September 1912): 436–39; *ToL*, May 29, 1914.
21. *Pensacola (FL) Journal*, November 24, 1906; *ToL*, March 10, 1910; *NYT*, June 27, 1915; *Washington Herald*, June 22, 1910; *New York Tribune*, April 2, 1906.
22. *ToL*, April 13, 1910; Dr. J. T. Allen, "Daily Diet Hints," *WP*, June 27, 1909.
23. The company was incorporated as the Franco-American Ferment Company. Metchnikoff emphasized that he received no remuneration for the endorsements. *NYT*, May 25, October 13, 1909; January 11, January 16, 1910; April 14, May 16, 1915; *New York Sun*, October 25, 1908.
24. A. J. Cramp, "The JBL Cascade Treatment," *JAMA* 63 (1912): 213.
25. *Washington Herald*, February 13, 1910; *ToL*, April 29, 1910; Cramp, "Cascade Treatment," 213; *NYT*, June 19, 1910; Edwin Slosson, *Major Prophets of Today* (Boston, 1916), 175, cited in Whorton, *Crusaders*, 220.
26. Metchnikoff said that his diet now centered on adding two kinds of "good microbes" into his system, one producing sugar and the other lactic acid. *ToL*, June 20, 1914.
27. Vaughan, "The Romantic Rationalist."
28. Whorton, *Inner Hygiene*.
29. Levenstein, *Revolution*, 21.
30. *NYT*, October 20, 1874. First diagnosed by an American, Dr. Abernathy, before the Civil War, dyspepsia was regarded as a quintessentially American disease, the result of "materialism and overwork." *NYT*, January 28, 1854. The most effective of the cures was probably "MAN-A-CEA, Natural Spring Water." *NYT*, May 21, 1900. The most famous of the cures was Carter's Little Liver Pills, an ineffective little pill that was a reliable moneymaker for over a hundred years, until the 1950s.

31. *NYT*, April 11, 1913; January 14, 1914; *ToL*, August 21, 1939.
32. *NYT*, April 9, 1907.
33. John Harvey Kellogg, *Autointoxication or Intestinal Toxemia* (Battle Creek, MI, 1919), 311, cited in Whorton, *Crusaders*, 221. Kellogg blamed constipation-induced autointoxication for everything from cancer to manic-depressive disorder and schizophrenia. When he suggested that the fiery populist William Jennings Bryan might be suffering from autointoxication, the puzzled Bryan asked, "Is that something one gets from driving too rapidly in an automobile?" Richard Schwarz, *John Harvey Kellogg, M.D.* (Nashville: Southern Publishing, 1970), 54.
34. There are a number of amusing works about Kellogg and his "San," the most enjoyable of which are Gerald Carson, *Cornflake Crusade* (New York: Rinehart, 1957), and T. C. Boyle's hilarious and horrifying novel, *The Road to Wellville* (New York: Viking, 1993).
35. It was Kellogg who actually came up with the term "Fletcherize." J. H. Kellogg to Horace Fletcher, August 21, September 3, 1903, in Fletcher, *New Glutton*, 67, 77; Levenstein, *Revolution*, 87–91; Whorton, *Crusaders*, 168–200.
36. Fletcher's ideas appealed to scientists arguing for reducing the current high estimates for protein requirements. His experiments showed that "chewing," which naturally led to reduced consumption of protein, had no adverse effects. Fletcher, *New Glutton*, 24–25; Levenstein, *Revolution*, 88–94.
37. Fletcher, *New Glutton*, 144–45, 174–75; Levenstein, *Revolution*, 88–90.
38. Kellogg also disagreed with Metchnikoff's idea that modern humans were saddled with out-of-date appendages, finding it impossible to accept the idea that God, who created perfect Nature, could endow man with useless organs like the colon. He thought the colon's problems were caused by "civilization." Whorton, *Crusaders*, 220.
39. Kellogg, *Autointoxication*, 311.
40. John Harvey also refused to allow his name, or that of the San, to be used for selling laxatives or laxative foods that purported to combat autointoxication.
41. Logan Clendening, "A Review of the Subject of Chronic Intestinal Stasis," *Interstate Medical Journal* 22 (1915): 1192–93, cited in Whorton *Crusaders*, 217–19.
42. Cramp, "Cascade Treatment," 213.
43. Clendening, "Review," in Whorton, *Crusaders*, 217; A. N. Donaldson, "Relation of Constipation to Intestinal Intoxication," *JAMA* 78 (1922): 884–88, cited in Ernst, "Colonic Irrigation," 196–98.
44. *ToL*, May 8, 1922; Bengamin [*sic*] Gayelord Hauser, *Diet Does It* (London: Faber and Faber, 1952), 14.
45. Ernst, "Colonic Irrigation," 198.
46. Review of Carl Ramus, *Outwitting Middle Age* (New York: Century, 1926), in *The Bookman: A Review of Books and Life* 64 (November 1926): 3.
47. Hillary Schwartz, *Never Satisfied: A Cultural History of Diets, Fantasy and Fat* (New York: Free Press, 1986), 200.
48. *NYT*, September 10, 1998; May 21, 2009; "Dannon Co., Inc., Company History," http://www.fundinguniverse.com/company-histories/Dannon-Co-Inc-Company-History.html.
49. This was likely a kind of arthritis caused by the tuberculosis bacterium, *Mycobacterium*

tuberculosis, which causes symptoms such as weakness and sweating along with joint pain.

50. Hauser was born Eugene Helmuth Hauser, but when he went into the health food business, he changed it to Bengamin (*sic*) Gayelord Hauser. The jazz musician Louis Armstrong swore by Swiss Kriss, which still sells well in the United States and in Europe. Morris Fishbein, "Modern Medical Charlatans: II," *Hygeia* 16 (January 1938): 113–14; Karen Swenson, *Greta Garbo: A Life Apart* (New York: Scribner, 1997), 393, 398.
51. Hauser later claimed to have been inspired by Mahatma Gandhi, who told him, "I eat innocent food . . . goat's milk, dates, yogurt and vegetables." *NYT*, March 24, 1974. Although he was a firm believer in the curative powers of vegetables, Hauser was not a vegetarian and prided himself in having converted Garbo from it. Swenson, *Garbo*, 392.
52. Bengamin Gayelord Hauser, *The New Health Cookery* (New York: Tempo, 1930), 316; Hauser, *Look Younger, Live Longer* (New York: Farrar, Straus, 1950); Hauser, *Diet Does It*, 315–16.
53. "Yogurt History," http://homecooking.about.com/library/weekly/aa031102a.htm.
54. *NYT*, July 18, 1963.
55. USDA, Economic Research Service, "U.S. per capita food consumption, Dairy (individual), 1909–2004"; Rebecca Williams, "Yogurt: The Curds and Whey to Health?" *FDA Consumer*, June 1992; *NYT*, May 1, 1992; "Dannon, Inc. Company History." Gervais Danone, the French conglomerate that purchased Dannon from Beatrice in 1982, was the product of a merger with the French Danone.
56. Dannon had periodically played on this, most notably in 1973, with a famous TV commercial featuring an eighty-nine-year-old native of then Soviet Republic of Georgia eating its yogurt while a voice-over said, "And this pleased his mother very much. She was 114." "Dannon, Inc., Company History."
57. www.stonyfieldfarm.com, May 3, 2004.
58. Cindy Hazen, "Cultured Dairy Products," *Food Product Design*, March 2004; *G&M*, March 26, 2008.
59. *G&M*, September 5, 2006.
60. http://danactive.com, December 5, 2008. Lactobacillus pills also made a comeback in orthodox medical circles. In 2008, on a hospital visit to a friend suffering from a serious case of *C. difficile*, which causes terrible diarrhea, I asked the nurse what kind of pill she was giving him. "Oh," she said, "it's a lactobacillus pill, which contains good bacteria that fights the bad bacteria in the colon."
61. *NYT*, September 20, 2009; University of California, Berkeley, *Wellness Letter*, 25 (September 2009): 4.
62. *Guardian Weekly* (London), October 16, 2009.

Chapter Four

1. Evan Jones, *American Food: The Gastronomic Story*, 2nd ed. (New York: Vintage, 1998), 123.
2. Mark Twain, *Innocents Abroad* (Hartford, CT: American Publishing Co., 1869), cited in Erich Segal, "Thoughts for Foodies," *Times Literary Supplement*, December 23–29, 1988.
3. Levenstein, *Revolution*, 21.

4. The hygienic conditions in the abattoirs often left something to be desired. In Manhattan the part of First Avenue now occupied by the United Nations complex housed a string of abattoirs with carcasses hanging to "cool" on the sidewalks. Half of the block north of there was a huge horse manure pile, whose contents frequently blew in the wind around them. Behind it was a cove whose water was polluted by run-off from the manure pile. On the other side of the manure pile were plants making oleomargarine out of tallow and brewing lager beer. *NYT*, May 11, 1884.
5. Herbert W. Conn, "Recent Contributions to Social Welfare. Bacteriology: Food, Drink, and Sewage," *Chautauquan* 40 (September 1904): 73–74.
6. Roger Horowitz, *Putting Meat on the American Table: Taste, Technology, Transformation* (Baltimore: Johns Hopkins University Press, 2005), 27–30.
7. James Harvey Young, *Pure Food: Securing the Federal Food and Drugs Act of 1906* (Princeton, NJ: Princeton University Press, 1988), 226.
8. The term "embalmed" had actually originated with an army surgeon who testified that he had seen fresh beef that had clearly been treated with chemicals and "had the odor of embalmed bodies." *NYT*, December 22, 1898.
9. *NYT*, May 9, 1899. The argument was not yet that the chemicals themselves were dangerous. For example, salicylic acid was recognized in the *U.S. Pharmacopeia* as useful in treating rheumatism and gout. Suzanne Rebecca White, "Chemistry and Controversy: Regulating the Use of Chemicals in Foods, 1880–1959" (PhD diss., Emory University, 1994), 13.
10. Unlike Miles, Roosevelt's troops did not find the fresh refrigerated beef offensive. He said that after cutting off the black crust surrounding it (which may have resulted from the borax and boric acid with which it was coated to prevent decay), it was eaten with some relish. Many men did get sick after eating it, he said, but he thought this was because by then they were unused to eating fresh meat. *NYT*, March 26, 1899.
11. Edward F. Keuchel, "Chemicals and Meat: The Embalmed Beef Scandal of the Spanish-American War," *Bulletin of the History of Medicine* 48 (1974): 251.
12. Ibid., 257–60; Young, *Pure Food*, 136–39.
13. *New York Tribune*, February 18, 1899, in Keuchel, "Chemicals," 262; *NYT*, July 5, July 19, September 25, 1900.
14. *NYT*, March 30, 1902.
15. The dedication to Sinclair's *The Jungle* reads, "To the Workingmen of America," and the book ends with its Lithuanian immigrant hero joining the socialist movement.
16. Upton Sinclair, *The Jungle* (New York: Norton, 2003), 95, 131–32. A year earlier, John Harvey Kellogg had written a vegetarian tract that described the food that came out of the filthy abattoirs similarly. "Each juicy morsel of meat," he said, "is fairly alive and swarming with the identical micro-organisms found in the dead rat in the closet of the putrefying carcass of a cow." John Harvey Kellogg, *Shall We Slay to Eat?* (Battle Creek, MI: Good Health, 1905), 107, in James Whorton, "Historical Development of Vegetarianism," *AJCN* 59 (May 1994): 1107.
17. Upton Sinclair, "What Life Means to Me," *Cosmopolitan*, October 1906, 594.
18. *NYT*, June 5, 1906.
19. Peter Finley Dunne, *Dissertations by Mr. Dooley* (New York: Harper, 1906), 249, in Rich-

ard Cummings, *The American and His Food: A History of Food Habits in the United States*, rev. ed. (Chicago: University of Chicago Press, 1941), 99.

20. Later *The Jungle* was said to have cut meat consumption in half, but the only evidence for this was the testimony of an anti-Sinclair meat packer at a congressional hearing on the meat inspection bill who claimed that sales of fresh and preserved meats had "apparently" been halved. *Conditions in the Stockyards, Hearings before the Committee of Agriculture on HR 18,537, 59 Cong. 1 sess.* (Washington, DC, 1906), 75, cited in Young, *Pure Food*, 231. Since statistics on national meat consumption were not collected until 1909, there is no reliable way of substantiating this claim. On the other hand, I found no mention of it in the press of the time. USDA, Economic Research Service, "U.S. per capita food consumption, Red Meat, 1909-2004."
21. *NYT*, May 29, June 15, 1906.
22. Their main complaint against the law was that it applied only to meat shipped across state lines, and small packers who dealt in only one state were exempt from its provisions. Levenstein, *Revolution*, 39; Horowitz, *Meat*, 59–60.
23. Levenstein, *Revolution*, 40–41.
24. *NYT*, May 30, 1906. Sinclair wrote that "it was the custom, whenever meat was so spoiled that it could not be used for anything else, either to can it or chop it up into sausage." Sinclair, *Jungle*, 131.
25. Horowitz, *Meat*, 32–33.
26. In sixteenth- and seventeenth-century France, for instance, meat pâtés were almost universally distrusted, but they remained popular because consumers had confidence in the individual producers and vendors. Madeleine Ferrières, *Sacred Cow, Mad Cow: A History of Food Fears*, trans. Jody Gladding (New York: Columbia University Press, 2006), chap. 8.
27. *NYT*, August 29, 1905; David Hogan, *Selling 'Em by the Sack: White Castle and the Creation of American Food* (New York: NYU Press, 1997); Paul Hirshorn and Steven Izenour, *White Towers* (Cambridge, MA: MIT Press, 1979), 1–5.
28. A typical 1905 recipe for frankfurters called for 150 pounds of meat trimmings, 20 pounds of which were knuckle meat and 40 pounds of which were fat, to be combined with 40 pounds of water, 9 pounds of corn flour (i.e., cornstarch), salt, sugar, saltpeter, seasonings, and "1½ pounds of color water." Horowitz, *Meat*, 86.
29. *NYT*, May 30, 1906.
30. Hogan, *Selling*, 17.
31. *WP*, January 23, 1910. Horace Fletcher, urging the women's clubs in Pittsburgh to continue their boycott, predicted that "meat will cease to be eaten in America in ten years' time." This would be all to the good, he said, because prolonged meat eating caused "autointoxication, which is akin to alcoholic poisoning, and just as fatal in its effects." *NYT*, February 3, 1910.
32. *WP*, January 23, 1910; USDA, Economic Research Service, "U.S. per capita food availability, Beef, 1909-2005."
33. Werner Sombart, *Why Is There No Socialism in America?*, trans. Patricia M. Hocking and P.T. Husbands (White Plains, NY: M. E. Sharpe, 1906), 106.
34. Levenstein, *Revolution*, 138–45.

35. "Well Done," *University of Minnesota Medical Bulletin*, Winter 2008; Hogan, *Selling*, 25–35.
36. White Tower, though, targeted the working-class market. Levenstein, *Paradox*, 50, 228–29.
37. Arthur Kallett and Frederick J. Schlink, 100,000,000 *Guinea Pigs: Dangers in Everyday Foods, Drugs, and Cosmetics* (New York: Vanguard, 1933), 38–45. This set off a minor wave of "guinea pig journalism" and was followed by Schlink's *Eat, Drink and Be Wary* (New York: Covici, Friede, 1935).
38. Charles Wertenbaker, "New York City's Haughtiest Eatery," *Saturday Evening Post*, December 27, 1952.
39. Amy Bentley, *Eating for Victory: Food Rationing and the Politics of Domesticity* (Urbana: University of Illinois Press, 1998), 35–37; Allen J. Matusow, *Farm Policies and Politics in the Truman Years* (Cambridge, MA: Harvard University Press, 1967), 49.
40. Levenstein, *Paradox*, 98–99.
41. Bentley, *Eating for Victory*, 111.
42. USDA, ERS, "Beef, 1909–2005."
43. *NYT*, July 16, 1969.
44. Levenstein, *Paradox*, 169–71.
45. USDA, ERS, "Beef: Supply and disappearance, 1909–2005"; "Beef: Per Capita Availability adjusted for loss, 1970–2007."
46. Michael Pollan, *The Omnivore's Dilemma: A Natural History of Four Meals* (New York: Penguin, 2006), 82–83.
47. *NYT*, January 9, 1994.
48. *NYT*, January 4, 1994.
49. *NYT*, February 6, 1993; January 6, October 12, 1994.
50. Even this timid effort aroused the ire of the meat and food industries. Their trade associations filed a suit to stop this, arguing that *E. coli* was no problem if meat was cooked well enough. *NYT*, November 20, 1994.
51. Judith Foulke, "How to Outsmart Dangerous E. Coli Strain," *FDA Consumer* 28 (January/February 1994): 7.
52. *NYT*, July 23, November 20, 1994. Another estimate, also by the *New York Times*, was four hundred deaths a year. *NYT*, April 12, 1995.
53. Three years later, the Centers for Disease Control said the rate of infection from the dangerous form of *E. coli* remained steady, with 73,000 people stricken and 61 killed a year. *WP*, April 9, 2001.
54. *NYT*, August 16, August 22, October 4, 1997.
55. Jeffrey Kruger and Dick Thompson, "Anatomy of an Outbreak," *Time*, August 3, 1998.
56. *NYT*, December 28, 1994; September 2, 1997.
57. *NYT*, March 31, 1996; Claude Fischler, "The Mad Cow Crisis in Global Perspective," in *Food and Global History*, ed. Raymond Grew (Boulder, CO: Westview, 1999): 207–13; Ferrières, *Sacred Cow*, 1–2.
58. From 1995 to 2007, a total of 200 human cases were reported worldwide, 164 of which were in the UK and 21 in France. Most of the others who died, including two of the three Americans, were among people who had lived in the UK when the disease was spreading there. U.S. Department of Health and Human Resources, Centers for Disease

Control and Prevention, "vCJD (Variant Creutzfeldt-Jakob Disease)," http://www.cdc.gov/ncidod/dvrd/vcjd/factsheet_nvcjd.htm; University of Edinburgh, the National Creutzfeldt-Jakob Disease Surveillance Unit, "CJD Statistics," http://www.cjd.ed.ac.uk/figures.htm.

59. *NYT*, December 3, 1997.
60. In the five years after the Jack in the Box outbreak, there were as many as thirty *E. coli* outbreaks a year, causing thousands of people to become very ill and hundreds to die. *NYT*, October 14, 1998; September 9, 1999; October 10, 2003; "Anatomy of an Outbreak," *Time*, August 3, 1998; *WP*, April 9, 2001.
61. Eric Schlosser, *Fast Food Nation: The Dark Side of the All-American Meal* (Boston: Houghton Mifflin, 2001). The book was subsequently made into a movie.
62. "U.S. Opposes JBS Plan to Buy National Beef," *Reuters*, October 20, 2008.
63. See, for example, Michael Pollan, "The Vegetable Industrial Complex," *NYTM*, October 15, 2006.
64. Pollan, *The Omnivore's Dilemma* and *In Defense of Food: An Eater's Manifesto* (New York: Penguin, 2008).
65. *NYT*, September 15, October 7, October 13, October 15, 2006. It was still down substantially two and a half years later. *NYT*, March 11, 2009.
66. *NYT*, October 31, 2006.
67. There was a simultaneous outbreak in a chain called Taco John's. Both were later traced to California lettuce. December 5, December 15, December 27, 2006; February 20, 2007.
68. *NYT*, September 27, September 30, October 1, 2007; "Topps Meat Company Voluntarily Recalls Ground Beef Products that May Contain E. Coli 057:H7," www.toppsmeat.com, September 26, 2007.
69. *NYT*, October 5, October 6, 2007.
70. *NYT*, February 18, 2008; "Authorities Investigate Big Rise in Beef Contamination," *Consumer Reports*, March 2008, 11.
71. USDA, Food Safety and Inspection Service, "Letter from Thermy," http://www.fsis.usda.gov/food_safety_education/letter_from_thermy.
72. Roth's company claimed that when this frozen product was added to regular ground beef, it killed any such pathogens in it as well. *NYT*, December 31, 2009.
73. *NYT*, December 31, 2009.
74. The food writer Mark Bittman adds concerns over the environmental degradation caused by raising beef, including grass-fed beef, and calls for not eating livestock of any kind. Mark Bittman, *Food Matters: A Guide to Conscious Eating with More than 75 Recipes* (New York: Simon and Schuster, 2009).
75. Nevertheless, Schlosser and Pollan did join in urging the bill's passage. *NYT*, November 29, November 30, 2010.

Chapter Five

1. John Duffy, *The Sanitarians: A History of American Public Health* (Urbana: University of Illinois Press, 1990), 186. As with the later Pure Food and Drugs Act, many of these were supported by producers seeking to suppress corner-cutting competitors. Mitch-

ell Okun, *Fair Play in the Marketplace: The First Battle for Pure Food and Drugs* (DeKalb: Northern Illinois University Press, 1986).

2. "Against Poison and Fraud," *Outlook* 83 (1906): 496–97, cited in James Harvey Young, *Pure Food: Securing the Federal Food and Drugs Act of 1906* (Princeton, NJ: Princeton University Press, 1988), 253.
3. Harvey W. Wiley, *Harvey W. Wiley: An Autobiography* (Indianapolis: Bobbs-Merrill, 1930), 50; Oscar O. Anderson, *The Health of a Nation: Harvey W. Wiley and the Fight for Pure Food* (Chicago: University of Chicago Press, 1958), 263.
4. Suzanne Rebecca White, "Chemistry and Controversy: Regulating the Use of Chemicals in Foods, 1883–1959" (PhD diss., Emory University, 1994), 2–4. In his autobiography, Wiley said that he resigned after being reprimanded for bringing the university into disrepute by riding one of the first bicycles to be seen in Lafayette, "dressed up like a monkey and astride a cartwheel." This may well be true. Wiley owned the second automobile in Washington, D.C., and was in its first serious auto accident. Wiley, *Autobiography*, 158, 279.
5. Ilyse D. Barkan, "Industry Invites Regulation: The Passage of the Pure Food and Drug Act of 1906," *AJPH* 75 (January 1985): 20–21; Edwin Björkman, "Our Debt to Dr. Wiley," *World's Work* 19 (January 1910): 124–46.
6. James Harvey Young, "The Science and Morals of Metabolism: Catsup and Benzoate of Soda," *Journal of the History of Medicine* 23 (January 1968): 86.
7. Wiley testified to a government inquiry that his department found no chemicals in the canned beef it examined and opined that the health problems were probably the result of the hot climate and poor handling, rather than any chemicals. Edward F. Keuchel, "Chemicals and Meat: The Embalmed Beef Scandal of the Spanish-American War," *Bulletin of the History of Medicine* 48 (1974): 258.
8. *NYT*, November 1, 1903.
9. Lorine Goodwin, *The Pure Food, Drink, and Drug Crusaders, 1879–1914* (Jefferson, NC: McFarland, 1999), 49.
10. Goodwin, *Pure Food*, 150–64.
11. *WP*, January 2, 1903. Actually, after it proved difficult for them to eat foods containing what Wiley had warned them were disgusting chemicals, he was forced to feed them the chemicals in capsule form. Young, "Science and Morals," 89.
12. *NYT*, May 22, 1904.
13. *NYT*, February 1, 1903; "Song of the Poison Squad," in U.S. National Archives Exhibit, "What's Cooking Uncle Sam?," Washington, DC, June 10, 2010–January 12, 2011; Young, *Pure Food*, 136–37; Wiley, *Autobiography*, 217–20.
14. *NYT*, September 18, 1904; Anderson, *Health of a Nation*, 152. A sign of the times was that the *New York Times* reported, without comment, that in testifying before a congressional committee in 1906, Wiley was asked by a congressman from Georgia, "Do you really conduct a boarding house on pills and paregoric?" "Yes sir," replied Wiley, "we would be glad to have you come around." The congressman replied, "Well. I think I'd rather have nigger women cook for me." *NYT*, February 27, 1906.
15. Björkman, "Our Debt to Dr. Wiley," 124–44.
16. At that time food processors accounted for one-fifth of the nation's total manufactur-

ing activity. James Harvey Young, "Food and Drug Enforcers in the 1920s: Restraining and Educating Business," *Business and Economic History* 21 (1992): 121.

17. *NYT*, November 8, 1901; November 1, 1903. This was ultimately perhaps the most important reason that the large producers came around to supporting the Pure Food and Drug Act of 1906. Barkan, "Industry Invites Regulation."
18. Richard O. Cummings, *The American and His Food: A History of Food Habits in the United States*, rev. ed. (Chicago: University of Chicago Press, 1941), 97–98. The difference between Wiley's and Sinclair's attitudes toward the giant food producers was exemplified during the great meat boycott of 1910. Sinclair blamed high prices on the monopolistic "Beef Trust," while Wiley blamed the smaller "meat dealers," who acted as middlemen. *WP*, January 23, 1910.
19. This was the Association of Manufacturers and Distributors of Food Products, composed of the thirty-six largest manufacturers of canned fruits, vegetables, jams, jellies, preserves, and condiments. *NYT*, November 1, 1903.
20. Levenstein, *Revolution*, 41; Goodwin, *Pure Food*, 6.
21. *WP*, December 28, 1904; *NYT*, December 28, 1904.
22. *WP*, January 20, 1907; *NYT*, May 10, June 30, 1908.
23. *NYT*, July 23, 1911.
24. "Pure Food and Drug Act, 1906," United States Statutes at Large, 59th Cong., Sess. I, Chap. 3915, pp. 768–72.
25. This orientation complemented that of the drug section of the act, which was aimed at the fraud-filled patent medicine industry. In a sensational series of articles that *Collier's* magazine ran in 1905, subsequently published as a book, the muckraking journalist Samuel Hopkins Adams, aided by Wiley, chronicled numerous cases of proprietary medicines containing poisonous substances. Samuel Hopkins Adams, *The Great American Fraud* (Chicago, 1906).
26. A Kansas City newspaper thanked Wiley profusely, saying that now Americans could defy "those caviling investigators who have discovered so many germs and corresponding diseases that life seems worth nothing . . . and existence from day to day is a gamble with long odds against us." *Kansas City Journal*, n.d., quoted in *WP*, April 18, 1905.
27. The most common chemical preservative food processors used was benzoate of soda, followed by salicylic acid and borax. Few, they claimed, used formaldehyde, which they acknowledged was potentially harmful. *NYT*, November 1, 1903.
28. *WP*, November 21, November 22, November 23, 1906.
29. Wiley used what became his fall-back position to recommend it be banned: that food could be preserved without its aid. *WP*, October 28, 1906.
30. Clayton A. Coppin and Jack High, *The Politics of Purity: Harvey Washington Wiley and the Origins of Federal Food Policy* (Ann Arbor: University of Michigan Press, 1999), 118.
31. Ibid., 123.
32. *WP*, June 23, 1904; White, "Chemistry," 10–12.
33. Coppin and High, *Politics of Purity*, 119.
34. *WP*, January 25, 1908.
35. Björkman, "Our Debt to Dr. Wiley."

36. Young, "Science and Morals," 93.
37. *NYT*, September 14, 1907.
38. Young, "Science and Morals," 93.
39. *WP*, May 7, 1906; White, "Chemistry," 83. The pro-sulfur forces were helped by the fact that Wiley, who was a bit of a Francophile, would not support banning the importation of French wines, which also contained sulfur. Coppin and High, *Politics of Purity*, 119–21.
40. It was actually a recipe gleaned from a Pennsylvania housewife, who discovered that increasing the amounts of vinegar and sugar in the product allowed it to remain unspoiled for four weeks after the bottle was opened. The product was usually called catsup, but Heinz labeled his version "Ketchup." Coppin and High, *Politics of Purity*, 125–27.
41. *NYT*, October 3, 1912.
42. *NYT*, January 25, 1909; June 30, 1908; *WP*, May 3, September 13, 1909.
43. *NYT*, January 24, 1909; Coppin and High, *Politics of Purity*, 123.
44. *WP*, August 25, 1909. Lucrezia Borgia, a member of the powerful Borgia family in Renaissance Rome, was reputed to have slipped a deadly poison from a hollowed-out ring she wore into the drinks of various men who earned her enmity.
45. *NYT*, August 27, 1909. The homeopaths had endorsed his views the previous year, after being persuaded that benzoate was being used to disguise spoiled food. *WP*, October 9, 1909; Björkman, "Our Debt to Dr. Wiley," 124–47.
46. *NYT*, August 27, 1909. A chemist at the University of Wisconsin testified in Congress that benzoic acid "has been used for centuries without harm and should be used as a preservative because it is less harmful than anything that could be found." There was more benzoic acid in almond-flavored ice cream than in half a barrel of crispy pickles, and people had been eating substantial quantities of it in the cranberries that accompanied their turkeys with no ill effects. It also had medicinal value in the treatment of kidney diseases. (Wiley critics made much of the fact that cranberries, a traditional folk cure, naturally contain high levels of benzoic acid.) *WP*, February 16, 1906; Coppin and High, *Politics of Purity*, 128–33.
47. *WP*, August 8, 1909. Chlorosis is a benign form of iron-deficiency anemia in adolescent girls. By neurosis Wiley likely meant not Freud's kind, but a nervous disease, such as neurasthenia, which was thought to afflict middle-class women whose nervous systems could not withstand the pressures of modern life.
48. *WP*, December 17, 1909; Coppin and High, *Politics of Purity*, 144–45.
49. *NYT*, July 11, July 14, 1908.
50. *WP*, November 3, 1907. Wiley's main objection was that most added pure alcohol to the blend, which was an act of misrepresentation, and that the singular name "whisky" implied that they were produced from a single mash of grain. This would have driven practically all whiskies, except for Kentucky bourbon and Jack Daniel's, off the market.
51. *WSJ*, August 19, 1911; *NYT*, August 22, 1911.
52. *WP*, July 22, 1908; *NYT*, January 26, 1909; Young, "Science and Morals," 102.
53. Wiley's German co-discoverer was Constantine Fahlberg, who was working under him at the time. *WP*, April 29, 1911.

54. *NYT*, September 4, October 10, 1911. Most canners had come to realize that the process of heating the canned foods and keeping them in a vacuum was enough in itself to preserve the contents.
55. *NYT*, July 15, August 19, September 16, 1911; Wiley, *Autobiography*, 238–39.
56. *NYT*, November 16, 1911; March 16, 1912.
57. "By their fruits," Wiley said of Wilson's two opponents, "shall you know them, and these fruits have been scabby, worm-eaten, and rotten at the core." *NYT*, October 3, 1912.
58. *WP*, September 18, 1912.
59. *WP*, September 24, 1912; *NYT*, October 12, 1912.
60. *NYT*, November 1, 1907; June 19, July 11, 1908; July 25, 1910; *WP*, June 30, July 25, 1910.
61. The Remsen Board did come out against saccharin after its members subjected Poison Squads of their students to the product. *NYT*, April 29, 1911.
62. *WP*, September 24, 1912. This was a direct blow at Wiley. Two years earlier, although he himself had exempted French wine from the sulfur ban (prompting charges that this was prompted by his personal taste for the product and connections with its shippers) he accused former president Roosevelt of having secured "executive immunity" for canned peas with sulfate of copper imported from France because of his friendship with the French ambassador. *WP*, January 16, 1910.
63. Honoré Willsie, "Canning and the Cost of Living," *Harper's Weekly*, November 1, 1913, 8–9. Alsberg did, however, continue Wiley's assault on the caffeine in Coca-Cola. The case was finally settled in 1917 when the company agreed to substantially reduce the amount of it in its product without admitting that it was at all harmful. White, "Chemistry," 149.
64. In the case at hand, the large millers were bleaching dark-colored flour with nitrogen peroxide. Although this was a mixture of two admittedly dangerous chemicals, nitric and nitrous acid, the government could not absolutely prove that these chemicals were harmful to humans in the quantities used. Cummings, *The American and His Food*, 102–3.
65. The botulism was naturally occurring and had nothing to do with additives. White, "Chemistry," 200; Young, "Food and Drug Enforcers ."
66. James Whorton, *Before Silent Spring: Pesticides and Public Health in Pre-DDT America* (Princeton, NJ: Princeton University Press, 1974), 114.
67. The name was changed to the Food, Drug, and Insecticide Organization in 1927 and Food and Drug Administration in 1930.
68. *GH*, July 1928, 108.
69. Levenstein, *Revolution*, 197–98; Levenstein, *Paradox*, 13–14.
70. "Dr. Wiley's Question-Box," *GH*, April 1928, 102.
71. *WP*, September 3, 1933; February 23, 1936. The FDA head of education, Ruth de Forest Lamb, then expanded this into a best-selling book called *American Chamber of Horrors* (New York: Farrar and Rinehart, 1936).
72. Kallet and Schlink also revived germophobia, quoting, of all people, the turn-of-the-century authors H. G. Wells and Julian Huxley as warning that "often meat and other foods [in stores] are badly infected with bacteria" because of dust and flies and, as a fillip, revived the (false) idea that tubercular cattle caused tuberculosis, warning that in

1930 nearly 1 million people in Massachusetts were exposed to milk from tubercular cattle. *NYT*, December 17, 1932; February 3, 1934; Arthur Kallet and Frederick J. Schlink, *100,000,000 Guinea Pigs: Dangers in Everyday Foods, Drugs, and Cosmetics* (New York: Vanguard, 1933), 38–45.

Chapter Six

1. Elmer V. McCollum, *The Newer Knowledge of Nutrition* (New York: Macmillan, 1918); Levenstein, *Revolution*, chaps. 4, 6, 11.
2. George Wolf and Kenneth J. Carpenter, "Early Work into the Vitamins: The Work of Wilhelm Stepp," *JN* 127 (July 1997): 1255.
3. Aaron J. Ihde and Stanley L. Becker, "Conflict of Concepts in Early Vitamin Studies," *Journal of the History of Biology* 4 (Spring 1971): 1–33; Naomi Aronson, "The Discovery of Resistance: Historical Accounts and Scientific Careers," *ISIS* 77 (1986): 630–46. "Vitamine" was a contraction of Funk's original name, "vital amine." "Amine" means nitrogen, and when the compounds were eventually purified and discovered to contain little nitrogen, the "e" was dropped.
4. The term "vitamania" seems to have been coined in 1942, by Robert M. Yoder, in "Vitamania," *Hygeia* 20 (April 1942): 264–65. Rima Apple emphasizes the importance of its scientific backing in its rise, in her *Vitamania* (New Brunswick, NJ: Rutgers University Press, 1996).
5. Harrow added, optimistically, "but tens of thousands of lives have been saved by the intelligent application of the newer knowledge." Van Buren Thorne, review of *Vitamines: Essential Food Factors*, by Benjamin Harrow, *NYT Book Review*, March 27, 1921. The main problem was adjusting to the idea that many of the foods that the New Nutritionists had dismissed as practically without food value were full of vitamins. These included fresh citrus fruits, green vegetables, and other fresh foods with high water content.
6. Thorne, review of Harrow, *Vitamines*.
7. The latter two were favored by the Englishman Frederick Gowland Hopkins, who received the (much-disputed) Nobel Prize for the discovery of vitamins in 1929, and the American researcher Graham Lusk. Elmer V. McCollum, "Nutrition as a Factor in Physical Development," *Annals of the American Academy of Political and Social Science* 98 (November 1921): 36; Alfred C. Reed, "Vitamins and Food Deficiency Diseases," *Scientific Monthly* 13 (July 1921): 70.
8. The subtitle of McCollum's book reflects this: *Newer Knowledge: The Use of Food for the Preservation of Vitality and Health.*
9. Aronson, "Discovery," 636; Ihde and Becker, "Conflict of Concepts," 27–28. Doing nutrition research on cows was not as low-status as it might sound. The founder of modern nutritional science, the German chemist Justus von Liebig, came up with his most important discoveries by studying the feeding of cows.
10. Walter Gratzer, *Terrors of the Table: The Curious History of Nutrition* (New York: Oxford University Press, 2005), 164–67; Aronson, "Discovery," 636–38; Ihde and Becker, "Conflict of Concepts," 27–28; Elmer V. McCollum and Nina Simmonds, *The Newer Knowledge of Nutrition*, 3rd ed. (New York: Macmillan, 1925), 25–36. Rats were better subjects

than cows because they had a much shorter life cycle and diets and digestive systems that were more like humans.

11. Aronson, "Discovery," 636–38; McCollum and Simmonds, *Newer Knowledge*, 3rd ed. (1925), 21–41; Elmer V. McCollum, *From Kansas Farm Boy to Scientist: The Autobiography of Elmer Verner McCollum* (Lawrence: University of Kansas Press, 1964), 147–51.
12. *NYT*, June 19, 1922; M. K. Wisehart, "What to Eat: Interview with E. V. McCollum," *American Magazine*, January 1923, 15; Aronson, "Discovery," 638.
13. Nina Simmonds, "Food Principles and a Balanced Diet," *AJN* 23 (April 1923): 541–45; *NYT*, August 26, 1923; *WP*, March 23, 1924; Levenstein, *Revolution*, 149.
14. James Harvey Young, *The Medical Messiahs* (Princeton, NJ: Princeton University Press, 1967), 335.
15. Levenstein, *Revolution*, 156–58.
16. *NYT*, June 19, 1922.
17. Apple, *Vitamania*, 24–27.
18. Richard D. Semba, "Vitamin A as 'Anti-Infective' Therapy, 1920–1940," *JN* 129 (April 1999): 783–92; Apple, *Vitamania*, 20–25.
19. In 1923 *American Magazine* ran a series on how notable businessmen achieved success that proved to be very popular among ambitious young men. The editor then sent a young reporter to interview Elmer McCollum for a follow-up piece on businessmen's diets and how to combat the weight gain and high blood pressure that resulted from their sedentary lives. Much to everyone's surprise, the outpouring of mail that the article evoked was mostly from women. Wisehart, "What to Eat," 14–15; McCollum, *From Kansas Farm Boy*, 186.
20. At that early stage, it was called "the water-soluble vitamin," rather than vitamin B, perhaps because "B" might be construed as being of secondary importance. *American Magazine*, January 1922, 35; *Literary Digest*, June 10, 1922, 67.
21. See, for example, *GH*, September 1928, 138; January 1935, 100; U.S., Federal Trade Commission, Stipulation no. 02180, Cease and Desist Agreement of Standard Brands, Inc., July 28, 1938.
22. Reed, "Vitamins and Food Deficiency Diseases," 70; *NYT*, April 26, 1923.
23. *Saturday Evening Post*, June 11, 1932.
24. *NYT*, October 17, 1920; March 20, 1921; *Literary Digest*, June 10, 1922, 67; Apple, *Vitamania*, 27.
25. McCollum and Simmonds, *Newer Knowledge*, 3rd ed., 562–66; *Newer Knowledge*, 4th ed., 532–33.
26. McCullom even trotted out Metchnikoff, saying that his theory that the lactic acid that milk produces in the stomach gets rid of the nasty bacteria that causes food to putrefy "has been shown by bacteriological methods to be sound." Wisehart, "What to Eat," 112; McCollum and Simmonds, *Newer Knowledge*, 3rd ed., 565. In 1920, however, he had said that Metchnikoff's assertions about the healthful properties of *bacillus Bulgaricus* were "not supported by the many studies which have been made by later observers." He ascribed "the excellent health of the aged populations of the Balkan states as well as other peoples who live largely on milk products" to milk, "and not the presence of bacteria it may contain." Elmer V. McCollum and Nina Simmonds, *The American Home Diet* (Detroit: Matthews, 1920), 67.

27. For a while, McCullom included fresh crisp vegetables as a third protective food, both for roughage and as a source of newly discovered vitamin C. McCollum and Simmonds, *Newer Knowledge*, 3rd ed., 562–66.
28. Eunice Fuller Barnard, "In Food, also, a New Fashion Is Here," *NYTM*, May 4, 1930, 6.
29. See my *Revolution at the Table*, 157–60.
30. McCullom made an exception of vitamin D, which came from cod liver oil and sunlight. William M. Johnson, "The Rise and Fall of Food Fads," *American Mercury* 28 (April 1933): 476.
31. New Nutritionists tended to dismiss these concerns, arguing that the bran and any of its accompanying nutrients were indigestible. *NYT*, June 16, 1907.
32. McCullom did not use the still unfamiliar term "vitamins" but called it "very poor in the substances which protect the body against the deficiency diseases, scurvy, beri-beri and xerophthalmia." McCollum and Simmonds, *American Home Diet*, 47–49.
33. Elmer V. McCollum, *The Newer Knowledge of Nutrition*, 2nd ed. (New York: Macmillan, 1923), 416.
34. McCollum and Simmonds, *Newer Knowledge*, 3rd ed., 131.
35. McCollum, *From Kansas Farm Boy*, 196; Elmer V. McCollum and Nina Simmonds, *Food, Nutrition and Health* (Baltimore: the authors, 1925), 55.
36. Elmer McCollum and J. Ernestine Becker, *Food, Nutrition and Health*, 3rd ed. (Baltimore: the authors, 1933), 61–62.
37. *WP*, August 4, 1932.
38. M. Daniel Tatkon, *The Great Vitamin Hoax* (New York: Collier-Macmillan, 1968), 28.
39. It also ignored Casimir Funk. Instead the prize went to the Englishman Sir Frederick Gowland Hopkins, and a Dutchman, Christian Eijkeman. *NYT*, November 1, 1929.
40. General Mills, "Outline of the Career of Marjorie Husted, [Betty Crocker]," Marjorie Husted Papers, Schlesinger Library, Radcliffe Institute, Cambridge, MA; *Food Field Reporter*, October 9, 1933; Frederick J. Schlink, *Eat, Drink and Be Wary* (New York: Covici, Friede, 1935), 12, 19, 222. The real irony—or rather, tragedy—is that while vitamania was sweeping a middle class who had little to worry about regarding vitamin deficiencies, Dr. Joseph Goldberger of the U.S. Public Health Service was showing that pellagra, a devastating disease that struck poor people in the rural South, was caused by a vitamin deficiency resulting from their diet based on refined cornmeal. Yet he was met with a stone wall of rejection. Correcting the problem would have meant having poor sharecroppers turn cotton acreage over to vegetables and other food crops, undermining the one-crop reign of King Cotton. It was not until the boll weevil infestation devastated the cotton economy in 1927 and 1928 that the government began recognizing what Goldberger had been saying for years. It began making milk available and encouraged growing crops that would supply the much-needed niacin, whose absence in the diet was pellagra's main cause. Still, pellagra continued to afflict many thousands of people well into the 1930s, until New Deal programs and the war finally altered the old dietary regime.
41. McCollum, *From Kansas Farm Boy*, 196–98.
42. It was said to result from insufficient alkalinity in the body, caused by an excess of acid in the bloodstream. L. J. Henderson, "Acidosis," *Science*, n.s. 46 (July 27, 1917): 73–83;

Martha Koehne, "To Reduce or Not to Reduce," *AJN* 25 (May 1925): 370–75; Fassett Edwards, MD, "The Acid Test," *Collier's*, July 19, 1930, 23.

43. McCollum and Becker, *Food, Nutrition and Health*, 3rd ed., 112, 115. There is no mention of acidosis in the 1925 edition of *Food, Nutrition and Health*. McCollum and Simmonds, *Food, Nutrition and Health*.
44. Dr. W. A. Evans, "How to Keep Well," *WP*, January 18, 1929; Josephine H. Kenyon, "Acidosis," *GH*, May 1931, 106.
45. *WP*, January 22, 1928; McCollum and Becker, *Food, Nutrition and Health*, 3rd ed., 112.
46. Barnard, "In Food," 6; Bertha M. Wood, "Acidosis and Its Therapeutics," *AJN*, June 1929, 681.
47. *WP*, March 20, 1925; *GH*, April 1928, 199.
48. McCollum and Becker, *Food, Nutrition and Health*, 3rd ed., 111; Russell W. Buntin, "Recent Developments in the Study of Dental Caries," *Science* 78 (November 10, 1933): 421. (Previously, too much fat in the diet was said to cause caries. Bertha Wood, "Measured Food," *AJN* 26 [February 1926]: 107.) Inis Weed Jones, "What Is Acidosis?" *Parents*, December 1934, 26; Kenneth Roberts, "An Inquiry into Diets," in *For Authors Only, and Other Gloomy Essays* (Garden City, NY: Doubleday, 1935), 89.
49. T. Swann Harding, "Common Food Fallacies," *Scientific Monthly* 25 (November 1927): 455–56; Mary Swartz Rose, "Belief in Magic," *JADA* 8 (1933): 494; James A. Tobey, "The Truth about Acidosis," *Hygeia*, August 1937, 693.
50. *AJN* 34 (March 1934): 305.
51. *WP*, August 4, 1932.
52. Welch's Grape Juice, however, held out until 1937, when the new Federal Trade Commission forced it to stop claiming that its "predigested grape sugar actually 'burns up' the ugly fat that masks your beauty" and worked to "correct" acidosis." U.S. Federal Trade Commission, Stipulation no. 01620, False and Misleading Advertising: Grape Juice, May 28, 1937.
53. McCollum had begun with a much more complex message, warning people to eat enough "mineral nutrients" to "maintain a proper state of neutrality in the blood and in the other body fluids. At least nine of the mineral substances are indispensable to life: five of them are basic [i.e., alkaline] and four acidic." *WP*, January 22, 1928. However, the simpler idea that proteins and carbohydrates should not be mixed in the stomach, which dates from the earliest days of the New Nutrition, was much easier to comprehend. In 1930 a well-known New York doctor signaled its revival when he explained to the *New York Times* that proteins and starches were "chemically and physiologically incompatible in the stomach." Barnard, "In Food," 7.
54. Roberts, "An Inquiry into Diets," 89.
55. Ford also inflicted this diet on the children in two schools he subsidized in Michigan and Massachusetts. Barnard, "In Food," 7.
56. Rose, "Belief," 494.
57. "The Wonders of Diet," *Fortune*, May 1936, 90; Morris Fishbein, "Modern Medical Charlatans: II," *Hygeia* 16 (January 1938): 113–14; Hillary Schwartz, *Never Satisfied: A Cultural History of Diets, Fantasy and Fat* (New York: Free Press, 1986), 200.
58. Harmke Kamminga, "'Axes to Grind': Popularizing the Science of Vitamins, 1920's

and 1930's," in *Food, Science, Policy and Regulation in the Twentieth Century*, ed. David F. Smith and Jim Phillips (London: Routledge, 2000), 94–95; Levenstein, *Paradox*, 3–9, 53–64.

Chapter Seven

1. Elmer V. McCollum, "The Contribution of Business to the Consumer through Research," *Journal of Home Economics* 26 (October 1934): 510–11; *NYT*, December 7, 1932; April 18, 1941.
2. *NYT*, November 30, 1938; Levenstein, *Paradox*, 19, 112.
3. *NYT*, December 28, 1941.
4. Vitamin sales rose 830 percent from 1929 to 1939, jumping 54 percent from 1937 to 1939. *NYT*, April 27, 1939; *WSJ*, November 25, 1941; W. H. Sebrell, "Nutritional Diseases in the United States," *JAMA* 115 (September 7, 1940): 851. The market was dominated by major "ethical" drug manufacturers, such as Eli Lilly and Merck, as well as Lever Brothers. *WSJ*, November 25, 1941.
5. Despite the AMA, many doctors began prescribing vitamins, accounting for an estimated one-quarter of vitamin sales in 1940. Sebrell, "Nutritional Diseases," 852.
6. *NYT*, May 27, 1941. These studies were deeply flawed, greatly exaggerating the extent of malnutrition. See my *Paradox*, 53–60.
7. The British, who gave their night fighter pilots vitamin A, in the form of carrots, played a major role in promoting the idea that it was essential for night vision. *WP*, December 7, 1941. Soon after the war began, the Germans—fearing that shortages of milk, meat, fresh vegetables, and other carriers of the vitamin might lead to widespread night blindness—began manufacturing and distributing large quantities of vitamin A. *ToL*, December 29, 1939. They also pushed the production and distribution of other vitamins, perhaps responding to Hitler's well-known obsession with healthy eating.
8. Tooth decay was still thought to originate in a lack of vitamin C or D. *NYT*, April 18, April 30, 1941; *WP*, July 10, 1941. Later, when questioned about whether selectors were being too choosy in this regard, President Roosevelt responded that they did not want men in the U.S. Army with false teeth. One of the reporters pointed out that General Grant had false teeth, and might have added George Washington too, but FDR did not respond. *NYT*, October 4, October 11, 1941.
9. "Vitamin Famine Prevalent in the United States," *SNL*, June 22, 1940, 395.
10. E. M. Koch, "Refinement of Food and Its Effect on Our Diet: Part 1," *Hygeia* 18 (1940): 620; ibid., "Part II," 703–4, 718.
11. *NYT*, December 30, 1941.
12. *NYT*, May 27, 1941.
13. Levenstein, *Paradox*, 65–66. For example, scientists pushing for a high requirement of thiamine estimated that it should be 0.45 mg. per 1,000 calories, so the RDA was set at .6 mg. Russell M. Wilder, "Vitamin B_1 (Thiamine)," *Annals of the American Academy of Political and Social Science* 225 (January 1943): 30.
14. As a result, the U.S. Army Air Corps distributed so much vitamin A to prevent night blindness and improve vision that civilian consumption had to be restricted. *WP*, February 10, 1942; *NYT*, February 10, 1942. The Germans bought into this as well, wasting

precious resources in boosting vitamin A production to head off night blindness. *ToL*, December 29, 1939.

15. *NYT*, May 26, 1941.
16. "Vitamin Famine Prevalent." Together with sugar, white bread accounted for an estimated 55 percent of average caloric intake. John D. Black, "The Social Milieu of Malnutrition," *Annals of the American Academy of Political Science* 225 (January 1943): 144. Perhaps coincidentally, the main perpetrators, the large milling companies, were concentrated around nearby Minneapolis.
17. The advantage of using the asylum patients was that it allowed the kind of control over their food intake that was impossible with people who went home each day after work. Three of them were sisters whose mental illnesses were quiescent enough for them to help with hospital housework. The diet was based on white flour, sugar, white raisins, white rice, corn starch, tapioca, hydrogenated fat, and as much candy as they wanted. R. D. Williams, H. L. Mason, and B. F. Smith, "Induced Vitamin B_1 Deficiency in Human Subjects," *Proceedings of the Staff Meetings, Mayo Clinic* 14 (December 13, 1939): 791.
18. "Observations on Induced Thiamin (Vitamin B_1) Deficiency in Man," *Archives of Internal Medicine* 60 (October 1940): 785–89. The authors admitted that because many high-protein foods that contained thiamine—such as meats, milk, and legumes—had to be eliminated from the diet, it was very high in carbohydrates. Wilder, "Vitamin B_1 (Thiamine)," 31. This may well have led to marked ups and downs in glucose levels in the subjects' blood.
19. "Vitamins for War," *JAMA* 115 (October 5, 1940): 1198; *NYT*, January 22, 1941.
20. George R. Cargill, "The Need for the Addition of Vitamin B_1 to Staple American Foods," *JAMA* 113 (December 9, 1939): 2146–51.
21. In October 1939 Hoffman-La Roche pharmaceuticals told food processors that "Mrs. American Housewife" was now insisting that food labels contain "specific declarations, so many units of vitamin B, so many of C. . . . She knows that these two vitamins especially may be destroyed or lost in modern processing," and that "medical experts" now supported their restoration. *Food Field Reporter*, October 2, 1939. In July 1940 the British government ordered that thiamine be added to white bread, but the measure was postponed, initially because of a shortage of the vitamin, and eventually abandoned. *ToL*, July 19, October 30, 1940.
22. *NYT*, September 8, 1940.
23. *NYT*, October 4, 1940.
24. Rima Apple, "Vitamins Will Win the War: Nutrition, Commerce and Patriotism in the United States during the Second World War," in *Food, Science, Policy and Regulation in the Twentieth Century*, ed. David F. Smith and Jim Phillips (New York: Routledge, 2000), 136.
25. *SNL*, February 8, 1941; *NYT*, April 9, 1941; *WP*, March 8, 1941; *New Yorker*, March 15, 1941, 14.
26. *SNL*, February 8, 1941; *NYT*, January 30, March 19, 1941.
27. *WP*, May 27, 1941; *NYT*, March 4, 1941.
28. *SNL*, April 12, 1941; *NYT*, April 24, July 12, 1941; *WP*, March 23, 1942.
29. Albert M. Potts, "Nutritional Problems of National Defense," *Science*, n.s., 93 (June 6, 1941): 539.

30. Waldemar Kaempffert, "What We Know about Vitamins," *NYTM*, May 3, 1942, 23.
31. *NYT*, June 17, 1941.
32. *NYT*, June 9, 1942; June 18, 1943.
33. Wallace was quoting a radio commentator on this. Henry A. Wallace, "Nutrition and National Defense," in U.S. Federal Security Agency, *Proceedings of the National Nutritional Conference for Defense*, May 26–28, 1941 (Washington, DC: USGPO, 1942), 37.
34. The thiamine group could hold their arms out for between forty-three minutes and two hours, while the others could do so for only thirteen to sixteen minutes. *Drug Topics*, January 6, 1941, in Apple, "Vitamins Will Win the War," 138.
35. *NYT*, March 27, April 1, 1941; March 10, 1942.
36. *NYT*, January 4, 1942.
37. George Gallup, *The Gallup Poll* (New York: Random House, 1972), I:310.
38. *WP*, September 20, 1942; Levenstein, *Paradox*, 69.
39. *NYT*, January 17, 1942; *New Yorker*, April 24, 1943, 9; Gallup Organization, "Gallup Brain" website, Gallup Poll no. 266, April 15, 1942, question 10g, http://brain.gallup.com/documents/questionnaire.aspx?STUDY=AIPO0266&p=3.
40. *NYT*, April 29, 1941; Russell Wilder, "The Quality of the Food Supply," memo attached to Cummings to Wilson, October 10, 1942, in U.S. Executive Office of the President, Office of the Coordinator of Health, Welfare, and Related Defense Activities, Nutrition Division, Records, RG 136, entry 218, box 31. This was a reversal of a stand against fortification that Wilder had taken just fourteen months earlier, when he had condemned "trying to improve on nature by putting vitamins into foods which do not naturally contain them." *NYT*, June 16, 1941.
41. The AMA would go no further than supporting the enrichment of bread and milk, and even opposed adding vitamins to baby food. "Shotgun Vitamins Rampant," *JAMA* 117 (October 21, 1941): 1447; "Council on Food and Nutrition," *JAMA* 121 (April 24, 1943): 1369–70.
42. Wilder tried to discount these studies on the grounds that, unlike the Mayo women, the subjects had been deprived of the vitamin for such a long time that their conditions had become chronic and incurable. Russell Wilder, "Nutritional Problems as Related to National Defence," *Journal of Digestive Diseases* 8 (July 1941): 244–45.
43. *WP*, July 4, 1943; *WSJ*, June 7, 1943; Levenstein, *Paradox*, 69.
44. *NYT*, April 22, 1937; March 15, 1940; September 12, November 15, 1941. Both the Hearst-owned International News Service and King Features ran enthusiastic reports, the latter saying that in a test run on fifty people by a doctor in Boston, "in every case there was a considerable darkening of the hair." *WP*, November 17, October 21, 1941.
45. *NYT*, November 13, 1941; *WP*, October 15, November 17, 1941; January 4, 1942.
46. The companies, American Home Products, which had just purchased Ansbacher's company, and Abbot Laboratories, reported that it cured sterility in a number of women, as well as some cases of skin disease, asthma, and constipation. *NYT*, April 3, 1942.
47. *NYT*, November 13, 1941; *WP*, December 17, 1942. It also had to compete with another B-complex vitamin, inositol, which was also called an "anti-gray-hair vitamin," and another supposedly anti-aging nutrient in the B-complex, pantothenic acid. *NYT*, September 9, 11, 1942; February 14, 1943.

48. Ansbacher did point out that he said that it worked in only two-thirds of cases and that his wife's hair had changed from gray to its original dark brown. *WP*, April 25, 1942.
49. *Science*, n.s. 95 (April 10, 1942): 10.
50. *NYT*, May 16, 1943; October 8, 1944; May 25, November 4, 1945.
51. *NYT*, February 14, 1943.
52. *NYT*, September 16, 1941; February 27, 1942.
53. *NYT*, November 23, 1941; April 5, 1942; March 8, 1944. The food producers who survived the vitamin pill onslaught best were canned spinach producers. They cashed in on the popularity of the *Popeye* cartoons, whose sailor hero, when caught in dangerous situations, derived extraordinary energy by sucking down canned spinach. The acrid taste of canned spinach is another one of my unpleasant childhood memories.
54. The only difference that they were able to discern after massaging the figures was a slightly increased incidence of dry hair and follicular hyperkeratosis, a common, and insignificant, skin condition. Henry Borsok et al., "Nutritional Status of Aircraft Workers in Southern California: IV. Effects of Vitamin Supplementation on Clinical, Instrumental, and Laboratory Findings, and Symptoms," *Milbank Memorial Fund Quarterly* 24 (April 1946): 99–185.
55. Bengamin Gayelord Hauser, *Look Younger, Live Longer* (New York: Farrar, Straus, 1950), 162–63. The FDA embarked on a long, complicated prosecution of the publishers for allowing the book to be sold along with a Hauser-endorsed brand of blackstrap molasses, whose vitamin B content, said the book, would extend life by five years. *NYT*, March 10, March 11, March 27, June 26, August 4, August 16, 1951.
56. J. W. Buchan, "America's Health: Fallacies, Beliefs, Practices," *FDA Consumer*, October 1972, 5.
57. Levenstein, *Paradox*, 169, 202.
58. National Institutes of Health, "State-of-the-Science Conference Statement: Multivitamins/Mineral Supplements and Chronic Disease Prevention, May 15–17, 2006." Of course, this had no effect on the steady stream of claims, which continue unabated, supported by evidence that, to laymen at least, often looks convincing.
59. "Vitamins 'Could Shorten Lifespan,'" BBC News, http://news.bbc.co.uk/2/hi/health/6399773.stm, February 28, 2007; *NYT*, November 20, 2008; February 17, 2009.
60. *Pace* McCollum, it seems to have done nothing for my teeth.

Chapter Eight

1. Graham also thought that using manure to fertilize crops was "unnatural." Stephen Nissenbaum, *Sex, Diet and Disability in Jacksonian America* (Westport, CT: Greenwood, 1980), 7; Kyla Wazana Tompkins, "Sylvester Graham's Imperial Defence," *Gastronomica: The Journal of Food and Culture* 9 (2009): 59.
2. Robert McCarrison, "Faulty Food in Relation to Gastro-Intestinal Disorder," *JAMA* 78 (January 7, 1922): 2–4.
3. Robert McCarrison, *Studies in Deficiency Disease* (London, 1921); H. M. Sinclair, *The Work of Sir Robert McCarrison*, with additional introductory essays by W. R. Aykroyd and E. V. McCollum (London: Faber and Faber, 1953); McCarrison, "Faulty Food," 2–4;

"Memorandum on malnutrition as a cause of physical inefficiency and ill-health among the masses in India" (1926), in Sinclair, *The Work of Sir Robert McCarrison*, 261; McCarrison, "A Good Diet and a Bad One: An Experimental Contrast," *Indian Journal of Medical Research* 14, no. 3 (1927): 649–54; McCarrison, *Nutrition and National Health: Being the Cantor Lectures delivered before the Royal Society of Arts, 1936* (London: Faber and Faber, 1944), 29–30.

4. In 1939 McCarrison described this as food that "is, for the most part, fresh from its source, little altered by preparation, and complete; and that, in the case of those based on agriculture, the natural cycle is complete. No chemical or substitution stage intervenes." McCarrison in "Medical Testament: Nutrition, Soil Fertility, and the National Health," County Palatine of Chester, Local Medical and Panel Committee, March 22, 1939, journeytoforever.org/farm_library/medtest/medtest.html. This meant that people whose diets were based on whole grains, legumes, milk products, and fresh fruits and green vegetables were much healthier than those "whose diets are made up for the most part of denatured foodstuffs." *ToL*, April 3, 1939.
5. Philip Conford, "The Myth of Neglect: Responses to the Early Organic Movement, 1930–1950," *Agricultural History Review* 50 (2002): 101; McCarrison, "Introduction to the Study of Deficiency Disease," *Oxford Medical Publications* (1921), in McCarrison, *Nutrition and Health*, 94.
6. Eleanor Perényi, "Apostle of the Compost Heap," *Saturday Evening Post*, July 16, 1966, 30–33.
7. Ibid.; Wade Greene, "Guru of the Organic Food Cult," *NYTM*, June 6, 1971, 60; Samuel Fromartz, *Organic, Inc.* (Orlando, FL: Harcourt, 2006), 18–20.
8. This was much more important, Rodale said, than the three other reasons that McCarrison had given for their good health: that they were breast-fed, generally abstained from alcohol, and got plenty of exercise. J. I. Rodale, *The Healthy Hunza* (Emmaus: Rodale, 1948), 19–20.
9. *NYT*, June 13, 1948.
10. Guy Theodore Wrench, *The Wheel of Health: The Sources of Long Life and Health among the Hunza* (Milwaukee: Lee Foundation for Nutritional Research, 1945; New York: Schocken, 1972; New York: Dover, 2006).
11. Rodale was taken more seriously at subsequent hearings on pesticides and fertilizers, when the committee recommended investigating his claims about the superiority of organic over chemical fertilizers. Suzanne Rebecca White, "Chemistry and Controversy: Regulating the Use of Chemicals in Foods, 1883–1959" (PhD diss., Emory University, 1994), 316–21, 330.
12. *WP*, April 10, 1958; Levenstein, *Paradox*, 133.
13. It was called the Delaney clause because officially it was an amendment to the Food and Drugs Act.
14. White, "Chemistry," 360–65.
15. Levenstein, *Paradox*, 135.
16. Rachel Carson, *Silent Spring* (Boston: Houghton Mifflin, 1962), 13; Levenstein, *Paradox*, 134.
17. White, "Chemistry," 393.

18. *NYT*, February 11, September 25, 1957.
19. Renée Taylor and Mulford J. Nobbs, *Hunza: The Himalayan Shangri-la* (El Monte, CA: Whitehorn, 1962), 54.
20. Alan E. Banik and Renée Taylor, *Hunza Land: The Fabulous Health and Youth Wonderland of the World* (Long Beach: Whitehorn, 1960); John H. Tobe, *Hunza: Adventures in a Land of Paradise* (Emmaus, PA: Rodale, 1960); Renée Taylor, *Hunza Health Secrets for Long Life and Happiness* (Englewood Cliffs, NJ: Prentice-Hall, 1964); Taylor and Nobbs, *Hunza*; Jay M. Hoffman, *Hunza: Ten Secrets of the World's Oldest and Healthiest People* (Groton, 1968); Shahid Hamid, S. *Karakuram Hunza: The Land of Just Enough* (Karachi: Ma'aref, 1979). In France a Swiss book, Ralph Bircher's *Hunsa: Das Volk das keine Krankheit kennt* (1942), was translated and published in 1943 and went through four more editions until 1955 as *Les Hounza; un peuple qui ignore la maladie*, trans. Gabrielle Godet (Paris: Attinger, 1943; 1955). Another German-language book with a similar title appeared in 1978: Hermann Schaeffer, *Hunza: Volk ohne Krankheit* [Hunza: People without Sickness] (Düsseldorf: Diederichs, 1978).
21. The doctor was stationed in nearby Afghanistan. Paul Dudley White, *My Life and Medicine: An Autobiographical Memoir* (Boston: Gambit, 1971), 239.
22. *JAMA*, February 25, 1961, 106.
23. John Clark, *Hunza: Lost Kingdom in the Himalayas* (New York: Funk and Wagnalls, 1956), 64–74.
24. Barbara Mons, *High Road to Hunza* (London: Faber and Faber, 1958), 105. One reason for the illiteracy and absence of records was that their language, Burashaski, is unwritten. Alexander Leaf, "Observations of a Peripatetic Gerontologist," *NT*, September/October 1973, 6.
25. Kinji Imanishi, ed., *Personality and Health in Hunza Valley* (Kyoto: Kyoto University, 1963), 7–14.
26. Jewel H. Henrickson, *Holiday in Hunza* (Washington, DC: Review and Herald Publishing, 1960), 109–10.
27. I have argued elsewhere that this kind of belief in the efficacy of natural foods was akin to belief in a system of magic—that is, it was an alternative paradigm to the reigning scientific wisdom of the day, using Keith Thomas's analysis of how when organized religion solidified its grip on Western culture in the fifteenth and sixteen centuries, older, alternative ways of achieving such things as good health were called magic. Keith Thomas, *Religion and the Decline of Magic* (London: Weidenfeld and Nicolson, 1971). I said that now that modern medical science was the dominant way of preventing and curing illness, belief in the efficacy of natural and organic foods represented a kind of magical alternative to the reigning science-based paradigm. Harvey Levenstein, "Santé-Bonheur," in *Manger Magique: Aliments sorciers, croyances comestibles*, ed. Claude Fischler (Paris: Autrement, 1994), 156–68.
28. Rodale, *Healthy Hunza*, 37; Tobe, *Hunza*, 625, 629–30.
29. Organic Gardening and Farming, *Encyclopedia of Organic Gardening* (Emmaus, PA: Rodale, 1959), 47; Taylor, *Hunza Health Secrets*, 76; Tobe, *Hunza*, 629.
30. A typical Rodale article began by quoting an article by a very prominent biologist in the prestigious journal *Science* and then proceeded to praise the findings of a dentist,

published by a zany health food outfit in Milwaukee, regarding the superior skeletal structure of the inhabitants of remote valleys in the Swiss Alps, Indians in northern Canada, Eskimos (as Inuit were then called), and Australian aborigines. J. I. Rodale, "How Does Nutrition Affect Our Health?" *Prevention*, June 1968, 70–75.

31. Greene, "Guru," 30, 54.
32. Howard A. Schneider and J. Timothy Hesia, "The Way It Is," *Nutrition Reviews* 31 (August 1973): 236.
33. *NYT*, April 30, 1969; Greene, "Guru," 31.
34. U.S. Senate, Select Committee on Nutrition and Human Needs, *Nutrition and Private Industry: Hearings*, 90th Cong., 2nd sess., and 91st Cong, 1st sess., 1969, 3956–61.
35. Schneider and Hesia, "The Way It Is," 236.
36. James S. Turner, *The Chemical Feast* (New York: Grossman, 1970); Frances Moore Lappé, *Diet for a Small Planet* (New York: Ballantine, 1971); Boston Women's Health Collective, *Our Bodies, Ourselves* (New York: Simon and Schuster, 1976), 108.
37. Victor Herbert and Stephen Barrett, *Vitamins and "Health" Foods: The Great American Hustle* (Philadelphia: Stickley, 1981), 105.
38. *NYT*, June 1, 1974.
39. Greene, "Guru," 55.
40. Dick Cavett, "When that Guy Died on My Show," May 3, 2007, http://opinionator.blogs.nytimes.com/2007/05/03/when-that-guy-died-on-my-show/.
41. Greene, "Guru," 68.
42. Robert Rodale, "J. I. Rodale's Greatest Contribution," *OGF*, September 1971, 33.
43. "Death Rides a Slow Bus in Hunza," *Prevention*, May 1974; "Doctors Follow One Fad after Another," *Prevention*, July 1974, 74–75.
44. Marco Bruno, "The FDA Wants You to Hand in Your Brain," *Prevention*, February 1974, 117. It also attacked the agency's denial that foods grown organically were nutritionally superior.
45. Alexander Leaf, "Every Day Is a Gift When You Are over 100," *National Geographic*, January 1973, 93; Leaf, "Getting Old," *Scientific American* 229 (September 1973): 44; Leaf, "Observations of a Peripatetic Gerontologist," *NT* 8 (September/October 1973): 4. Five years later, Leaf repudiated these articles, saying that his caveats had been ignored and they had been misinterpreted. Leaf, "Paradise Lost," *NT* 13 (May/June 1978): 6–8.
46. *WP*, April 12, 1973.
47. Warren J. Belasco, *Appetite for Change: How the Counterculture Took on the Food Industry* (Ithaca, NY: Cornell University Press, 2006), 48–50.
48. When Rodale returned there in 1990 and saw chemical fertilizers and Kentucky Fried Chicken sweeping the country, he thought it was a disaster area. Anthony Rodale, "Robert Rodale: A Retrospective, 1930–1990," *Seeds of Change*, http://www.seedsofchange.com/cutting_edge/robert_rodale.aspx.
49. Robert Rodale, "Pleasures of the Primitive," *Prevention*, June 1974, 21–25; "Building Health with Chinese Herbs," *Prevention*, February 1974, 19–24; Michael Clark, "Jethro Kloss: Healing with Nature," *Prevention*, February 1974, 138–44.
50. Robert Rodale, "Gardening for Security," *OGF*, February 1975, 47.
51. Nevertheless, organic foods themselves were still considered fads in much of the main-

stream. See "Food Faddism Spurts as Young, Old People Shift to Organic Diets," *WSJ*, January 21, 1971.

52. *WP*, September 23, 1971; March 15, May 10, 1973; *NYT*, May 24, 1978.
53. Levenstein, *Paradox*, 195–96; *NYT*, March 15, 1978.
54. *Prevention* cited *Family Practice News*, March 15, 1974. "Bioflavonoids: Mother Nature's Answer to Female Problems," *Prevention*, October 1974, 86–87. Another typical one touted zinc supplements as "A New Therapy for Erectile Failure." It was based almost entirely on an article in a short-lived journal (*Sexual Medicine Today*) whose claim that zinc cured flagging penises was based on two studies saying that fifteen undernourished Egyptian boys and ten Iranian ones whose diets were supplemented with zinc, iron, and a number of other minerals matured sexually earlier than others. Mark Bricklin, "Zinc Makes You Think," *Prevention*, May 1974, 28–32.
55. That year, for example, a *Prevention* article advising that yogurt reduced vaginal yeast infections was based on a Long Island doctor's study of fifteen of his patients. There was no reference to yogurt-eating women in Bulgaria being free of the ailment. "Yogurt vs. Yeast," *Prevention*, May 1990, 20–23; Rebecca Williams, "Yogurt: The Curds and Whey to Health?" *FDA Consumer*, June 1992.
56. The large male readership of these magazines offered a unique attraction to advertisers, as health and fitness readers are usually overwhelmingly female.
57. *NYT*, October 12, 2003; Meghan Hamill, "Prevention Integrates Marketing to Grow Franchise," *Circulation Management*, March 25, 2005.
58. "Rodale Reports Second Quarter 2007 Results," http://www.rodale.com/. In 2009 it published a book critical of food processors by David Kessler, the ex-head of its historic *bête noire*, the FDA, but it dealt with the health consequences of obesity, something that was never high on the Rodales' hit list.
59. *NYT*, January 27, 2010.
60. Arlene Leonhard-Spark, "A Content Analysis of Food Ads Appearing in Women's Consumer Magazines in 1965 and 1977" (EdD diss., Teachers College, Columbia University, 1980), 989.
61. Eric Schlosser, *Fast Food Nation: The Dark Side of the All-American Meal* (Boston: Houghton Mifflin, 2001), 120–29. In 2004 Cadbury Schweppes advertised a new soda, 7Up Plus, as "100 per cent natural." After complaints that it was made with high-fructose corn syrup, the current villain of the day, it changed the label to "100 per cent natural flavor." *NYT*, March 7, 2007. Sixty-two percent of consumers also said the term "whole grain" would make them more likely to buy a product. Gallup Organization, "Gallup Brain" website, September Wave 4, September 26–29, 1991, question 43j, http://brain.gallup.com/documents/questionnaire.aspx?STUDY=AIPO0266&p=3.
62. *Tufts University Health and Nutrition Newsletter*, August 2008, 4–5; March 2009, 3; Marion Nestle, *What to Eat* (New York: North Point Press, 2006).
63. "Rodale Reports," 2007; Kraft website 2007, http://www.kraft.com/brands/namerica/us.html.
64. Belasco, *Appetite*, 185–242; Levenstein, *Paradox*, 195–212.
65. Michael Pollan, *The Omnivore's Dilemma: A Natural History of Four Meals* (New York: Penguin, 2006), 135–84.

66. See "When It Pays to Buy Organic," *Consumer Reports*, February 2006, 12–17; Michael Pollan, "Mass Natural," *NYTM*, June 4, 2006, 15–18.
67. Among those perpetuating it was Paavo Airola. A Finnish-born naturopath with a PhD in biology, he moved to Arizona and built a lucrative career selling books and lecturing to medical schools on how Americans should adopt the low-protein diet of the Hunzas, "the healthiest people in the world," which he said included yogurt made from raw milk. "The Pure, the Impure, and the Paranoid," *Psychology Today* 12 (October 1978), 68. He also warned that acidosis was the cause of many illnesses and recommended a diet with a proper acid/alkaline balance. Paavo O. Airola, *How to Get Well* (Phoenix: Health Plus Publishers, 1974). His idea that eating dried seaweed could prevent baldness was not nearly as persuasive, and his last book, *Worldwide Secrets of Living Young* (Phoenix: Health Plus Publishers, 1982), lost some of its credibility when he died of a stroke, at age sixty-four, a year after its publication.
68. Claude Fischler sent me photos of the display. The book was Jay M. Hoffman's *Hunza: Fifteen Secrets of the World's Healthiest and Oldest People* (New Win, 1997). It is a 1985 revised edition (with five more secrets) of his 1968 book, *Hunza: Ten Secrets of the World's Healthiest and Oldest People*. Rodale would not have gone along with the book's vegetarianism. He was a confirmed meat eater who raised his own beef cattle in Emmaus.

Chapter Nine

1. Claude Fischler seems to have coined the term "lipophobia," first in French, then in English. Claude Fischler, *L'Homnivore: Le goût, la cuisine et le corps* (Paris: Odile Jacob, 1990), 297; "From Lipophilia to Lipophobia," in *Dietary Fats*, by D. J. Mela (London: Elsevier, 1992).
2. I have argued that better diets played a major role in this. Levenstein, *Revolution*, 135. Gary Taubes described the impact on heart disease deaths aptly: "The actual risk of dying from a heart attack at any particular age remained unchanged. Rather, the rising number of 50-year-olds dropping dead of heart attacks was primarily due to the rising number of 50-year-olds." Gary Taubes, "The Soft Science of Dietary Fat," *Science*, March 30, 2001, 1.
3. Changes in how deaths were categorized in the early 1940s brought about one significant increase. Then another revision of the criteria from 1948 to 1949 led to the raising of the coronary death rates by about 20 percent for white males and about 35 percent for white females. Russell L. Smith and Edward R. Pinckney, *Diet, Blood Cholesterol and Coronary Heart Disease: A Critical Review of the Literature* (Sherman Oaks, CA: Vector Enterprises, 1991), vol. 2, 3–8.
4. Association of Schools for Public Health, "Health Revolutionary: The Life and Work of Ancel Keys," movie script, 2001, http://www.asph.org/movies/keys.pdf; *WP*, November 24, 2004; Ancel Keys, "Recollections of Pioneers in Nutrition: From Starvation to Cholesterol," *Journal of the American College of Nutrition* 9 (1990): 288–91; Keys, letter to the editor, *Time*, February 3, 1961.
5. The study's main conclusions, published in 1950, were that people recovering from starvation should eat more than the normal amount of calories and protein and be

given vitamin supplements. Some thought that the study's results did not justify the human suffering involved, but it was later welcomed as something that would have been impossible to do after international scientific organizations prohibited such studies as unethical. Ancel Keys, *The Biology of Human Starvation* (Minneapolis: University of Minnesota Press, 1950); William Hoffman, "Meet Mr. Cholesterol," *University of Minnesota Update*, Winter 1979; Theodore B. VanItallie, "Ancel Keys: A Tribute," *Nutrition and Metabolism* 2 (February 14, 2005): 4; Leah M. Kalm and Richard D. Semba, "They Starved So that Others Be Better Fed: Remembering Ancel Keys and the Minnesota Experiment," *JN* 135 (June 2005): 1347–52. See Todd Tucker, *The Great Starvation Experiment: The Heroic Men Who Starved So that Millions Could Live* (New York: Free Press, 2006), for a very positive account of the study.

6. In January 1948 Keys announced that his lab had discovered that along with thiamine deficiency's well-known propensity to lead to "fatiguability and lack of ambition," a deficiency of it, as well as of riboflavin and niacin, caused "psychoneuroses," with symptoms such as irritability, moodiness, and "mental depression." *WP*, January 4, 1954. Wilder had previously written an enthusiastic review of his work on starvation. Ancel Keys, *Adventures of a Medical Scientist* (New York: Crown, 1999), 36.
7. G. Lyman Duff and Gardner C. McMillan, "Pathology of Atherosclerosis," *American Journal of Medicine* 11 (July 1951): 92–108; Association of Schools for Public Health, "Health Revolutionary"; Frederick Epstein, "Cardiovascular Disease Epidemiology: A Journey from the Past into the Future," *Circulation* 93 (1996): 1755–64; Ancel Keys, Henry Longstreet Taylor, Henry Blackburn et al., "Coronary Heart Disease among Minnesota Business and Professional Men Followed Fifteen Years," *Circulation* 28 (1963): 381–95.
8. Keys, *Adventures*, 44; Keys, "Recollections"; Hoffman, "Mr. Cholesterol"; Ancel Keys and Margaret Keys, *How to Eat Well and Stay Well the Mediterranean Way* (Garden City, NY: Doubleday, 1975), 2. Claude Fischler originally pointed out this passage's similarity with Utopian literature. He also found similarities between belief in Keys's Mediterranean diet, which will be discussed in the next chapter, and my ideas about the magical nature of belief in the Hunza diet. Claude Fischler, "Pensée magique et utopie dans la science: De l'incorporation à la 'diète mediterranéenne,'" in *Pensée magique et alimentation aujourd'hui* (Paris: Les Cahiers de l'OCHA, 1996); Harvey Levenstein, "Santé-Bonheur," in *Manger Magique: Aliments sorciers, croyances comestibles*, ed. Claude Fischler (Paris: Autrement, 1994), 156–68.
9. *WP*, September 29, 1953; A. Keys, J. T. Anderson, F. Fidanza et al., "Effects of Diet on Blood Lipids in Man, Particularly Cholesterol and Lipoproteins," *Clinical Chemistry* 1 (1955): 34–52; Association of Schools for Public Health, "Health Revolutionary"; Keys, "Recollections," 289. In these early studies, Keys emphasized that it was not high-cholesterol foods, such as eggs, that caused high-serum cholesterol, but only high-fat foods. Smith and Pinckney, *Diet, Blood Cholesterol and Coronary Heart Disease*, vol. 2, 2–6.
10. Henry Blackburn, interview in Association of Schools for Public Health, "Health Revolutionary."
11. *WP*, September 15, 1954.
12. Paul Dudley White, *My Life and Medicine: An Autobiographical Memoir* (Boston: Gambit,

1971), 55–56; White, foreword to Ancel Keys and Margaret Keys, *Eat Well and Stay Well* (Garden City, NY: Doubleday, 1959), 7.

13. Previously, the society was "an elite club that held small meetings in pleasant places." Henry Blackburn, "Ancel Keys, Pioneer," *Circulation* 84 (1991): 1402–4.
14. William Hoffman, "Meet Monsieur Cholesterol," *University of Minnesota Update*, Winter 1979; *NYT*, September 14, 1954; *Circulation Research* 2 (1954): 392.
15. *Time*, January 31, 1955.
16. "Fat's the Villain," *Newsweek*, September 27, 1954; *NYT*, September 19, 1954.
17. *NYT*, September 26, 1955; Paul Dudley White, "White Links Coronary and World Peace," *WP*, October 30, 1955; "Heart Ills and the Presidency: Dr. White's Views," *NYT*, October 30, 1955; White, *My Life*, 181.
18. "The Specialized Nubbin," *Time*, October 31, 1955.
19. *WP*, October 13, 1955; Keys, *Adventures*, 62–69; "Capsules," *Time*, December 30, 1957; Keys and Keys, *Eat Well*, 6.
20. J. Yerushalmi and H. E. Hilleboe, "Fat in the Diet and Mortality from Heart Disease: A Methodological Note," *New York State Journal of Medicine* 57 (1957): 2343–54.
21. George V. Mann, "Diet-Heart: End of an Era," *NEJM* 297 (September 22, 1977): 644.
22. *NYT*, July 16, 1957; Yerushalmi and Hilleboe, "Fat in the Diet," 2343–54; Uffe Ravnskov, "A Hypothesis Out-of-Date: The Diet-Heart Idea," *Journal of Clinical Epidemiology* 55 (2002): 1057; *NYT*, November 25, 1956.
23. Stephen P. Strickland, *Politics, Science and Dread Disease* (Cambridge, MA: Harvard University Press, 1972), 83, 104, 142, 146–47, 21–23; *WP*, October 13, 1955. As part of the 1948 legislation, the name was changed to the National Institutes of Health.
24. "The Fat of the Land," *Time*, January 13, 1961; Keys, *Adventures*, 91; Henry Blackburn, "The Seven Countries Study: A Historic Adventure in Science," in *Lessons for Science in the Seven Countries Study*, ed. H. Toshima, Y. Koga and H. Blackburn (Tokyo: Springer-Verlag, 1994), 10–11. Keys's colleague Henry Blackburn later said, "White was essential to the realization of the ideas of Keys." Henry Blackburn, "Preventing Heart Attack and Stroke: A History of Cardiovascular Epidemiology. PDW: At the Center of It All," www.epi.umn.edu/cvdepi/essay.asp?id=95.
25. Ancel Keys, "A Brief Personal History of the Seven Countries Study," in *Lessons for Science*, ed. Toshima, Koga, and Blackburn, 3.
26. Ancel Keys, "Your Energy and Your Diet," *Vogue*, February 15, 1957; Keys, *Adventures*, 100; Keys and Keys, *Eat Well*, 59. Nowhere did Keys mention a National Dairy Council–funded study that he had published in 1950 showing that it was almost impossible to reduce serum cholesterol through dieting. It concluded that only a stark, almost completely cholesterol-free diet could do this. Ancel Keys et al., "The Relation in Man between Cholesterol Levels in Diet and in the Blood," *Science* 112 (July 21, 1950): 79–81.
27. White, preface to Keys and Keys, *Eat Well*, 8.
28. Keys and Keys, *Eat Well*, 53–59; "The Specialized Nubbin."
29. Keys and Keys, *Eat Well*, 284–85. Keys thought that the "preformed" cholesterol in egg yolks and organ meats was not, like that in saturated fats, absorbed into the blood. "Fats and Facts," *Time*, March 30, 1959.
30. Margaret Keys did reproduce a Neapolitan neighbor's recipe for making the pasta for

cannelloni by hand, but called for stuffing them with nonfat cottage cheese, buttermilk, and an egg. Keys and Keys, *Eat Well*, 219, 274.

31. *NYT*, April 5, 1959; "Fats and Facts."
32. *NYT*, May 1, May 3, November 1, 1959.
33. Keys, *Adventures*, 100.
34. It its fine print, the report said that fat reduction would mainly benefit people who were overweight, had already had a heart attack, or had a family history of heart disease, but the recommendations were still broadcast as applying to everyone. *NYT*, December 11, December 18, 1960; *WSJ*, December 12, 1960; "Fat in the Fire," *Time*, December 26, 1960.
35. *NYT*, December 15, 1960.
36. "Fat in the Fire"; *NYT*, December 12, 1960.
37. *WSJ*, December 13, 1960.
38. "Fat in the Fire."
39. *NYT*, August 7, 1962.
40. A very influential book by the historian David Potter argued that the American national character was the product of a long history of living in abundance. David Potter, *People of Plenty: Economic Abundance and the American Character* (Chicago: University of Chicago Press, 1954). The title of my book *Paradox of Plenty* is a riff on that.
41. *NYT*, September 27, 1959.
42. *NYT*, February 9, 1947; Peter Steincrohn, *What You Can Do for Angina Pectoris and Coronary Occlusion* (Garden City, NY: Doubleday, 1946).
43. "The Specialized Nubbin."
44. The idea that atherosclerosis, the main cause of heart disease, is a disease of modern civilization received quite a blow when an examination of ancient Egyptian mummies revealed that it afflicted seven of the eight men who died at over the age of forty-five. *G&M*, November 20, 2009.
45. O. W. Portman and D. M. Hegsted, "Nutrition," *Annual Review of Biochemistry* 26 (1957): 307–26, in Kenneth J. Carpenter, "A Short History of Nutritional Science," *JN* 133 (November 2003): 3331–42.
46. "Fat in the Fire"; *NYT*, September 19, 1954; Indeed, said a longtime colleague, "Keys's greatest hobby over the years was to poke fun at the insurance industry's equating relative body weight with excess risk of death." He said that Keys "delighted in demonstrating that . . . in most industrial countries those people in the middle range of body weight are far better off than those at either extreme." He also found that men who put on weight between the ages of forty and sixty had lower death rates than those who did not, attributing this in part to the fact that people who give up smoking put on weight. Henry Blackburn, "Ancel Keys," in "Minnesota Firsts," http://mbbnet.umn.edu/firsts/blackburn_h.html.
47. Keys and Keys, *Eat Well*, 20.
48. Keys and Keys, *How to Eat Well*; Fischler, "Pensée magique."
49. *NYT*, October 21, 1960; *WSJ*, December 12, 1960.
50. Keys and Keys, *Eat Well*, 56–57.
51. Gary Taubes, *Good Calories, Bad Calories* (New York: Knopf, 2007), 16.

Chapter Ten

1. Michael Pollan used "Our National Eating Disorder" as the title for the introductory chapter to *The Omnivore's Dilemma: A Natural History of Four Meals* (New York: Penguin, 2006), 1–11; David Steinberg, *The Cholesterol Wars* (Oxford: Academic Press, 2007).
2. Daniel Steinberg, "An Interpretive History of the Cholesterol Controversy: Part II: The Early Evidence Linking Hypercholesterolemia to Coronary Disease in Humans," *Journal of Lipid Research* 46 (February 2005): 179. Gary Taubes has an excellent analysis of the weaknesses of Keys's theory in *Good Calories, Bad Calories* (New York: Knopf, 2007): 4–41.
3. *WP*, December 29, 1965; January 11, 1970.
4. *NYT*, July 7, 1965.
5. *NYT*, October 9, October 10, 1946; February 3, February 9, June 7, 1947; February 3, 1951; April 5, 1953; American Heart Association, "History of the American Heart Association," AHA website, http://www.heart.org/HEARTORG/; Taubes, *Good Calories*, 9.
6. *NYT*, September 20, 1955.
7. *NYT*, October 9, October 10, 1946; February 9, 1947; April 5, 1953; *WP*, September 13, September 18, 1954.
8. Support for research rose from about one-quarter of the American Heart Association's budget in the late 1940s to about half in 1954. In 1955 the AHA boosted its research funding by 50 percent over the previous year, to close to $1.5 million. *NYT*, February 26, April 2, 1954; *Circulation Research* 3 (1955): 426.
9. *NYT*, January 18, 1960.
10. *NYT*, October 25, 1961.
11. *WSJ*, June 10, 1964.
12. *WSJ*, October 27, 1961; June 22, 1967; Irving H. Page and Helen B. Brown, "Some Observations on the National Diet-Heart Study," *Circulation* 37 (1968): 313–15. Nor was another study—funded by the U.S. Public Health Service, which aimed to enlist 100,000 men and women—completed. Leonard Engel, "Cholesterol: Guilty or Innocent?" *NYTM*, May 12, 1963.
13. The AHA report recommended that "as an experimental therapeutic procedure," physicians should try treating hardening of the arteries by telling patients to eliminate whole milk and cream from their diets, limit eggs to four a week, and to substitute lean meats, poultry and fish for fattier meats. "Vegetable oils should replace butter and lard in cooking." *NYT*, December 11, 1960.
14. *NYT*, March 4, August 7, 1962; *WSJ*, March 28, 1963; "Cholesterol Controversy," *Time*, July 13, 1962. Years later safflower oil was condemned for containing omega-6 fats, which were said to cause heart disease by promoting inflammation. *G&M*, April 8, 2009.
15. *WP*, June 14, 1971; Morton Mintz, "Eat, Drink and Be Merry," *WP*, March 18, 1973; R. L. Smith and E. R. Pinckney, *The Cholesterol Conspiracy* (St. Louis: Green, 1991), 125.
16. *NYT*, October 12, 1962.
17. George V. Mann, "Diet-Heart: End of an Era," *NEJM*, 297 (September 22, 1977): 646.
18. *NYT*, December 11, 1959.

19. *WSJ*, September 28, 1964.
20. The FDA also forced manufacturers of cholesterol-lowing drugs to tell physicians that it was not known if they would have "a detrimental, or beneficial, or no effect," on heart disease. *WSJ*, June 9, 1964; *NYT*, January 14, 1971.
21. *WP*, March 1, 1971; August 20, 1972.
22. *WP*, August 20, 1972; April 14, May 27, 1973; Mintz, "Eat, Drink."
23. Mann, "Diet-Heart," 646.
24. "Diet and Coronary Heart Disease," *Nutrition Reviews*, October 1972, 223.
25. Richard Podell, "Cholesterol and the Law," *Circulation* 48 (August 1973): 225–28; *WP*, July 15, 1976.
26. *WP*, May 27, 1973; Edward R. Pinckney and Cathey Pinckney, *The Cholesterol Controversy* (Los Angeles: Shelburne, 1973), 1.
27. It recommended that polyunsaturated fat intake be raised from the current 4 to 6 percent of the calories in diets to 10 percent. *WP*, August 20, 1972; *NYT*, October 25, 1974.
28. A major source of research funding was the recently founded American Health Foundation, whose chairman was the head of the company that made Wesson Oil and whose committee on food and nutrition was stacked with scientists working for it and other vegetable oil companies. In November 1971 it called on Congress to make increased use of polyunsaturated fats to reduce heart disease a "national policy." Mintz, "Eat, Drink." It subsequently shifted its target to cancer, funding research to prove that high-fat, low-fiber diets caused cancer.
29. *Food Processing*, April 1964, 65; Sheila Harty, *Hucksters in the Classroom: A Review of Industry Propaganda in the Schools* (Washington, DC: Center for the Study of Responsive Law, 1979), 23.
30. John Yudkin, *This Slimming Business* (London: MacGibbon & Kee, 1958). The book sold over 200,000 copies, a huge number for Great Britain. Victoria Brittain, "Not So Sweet on Sugar," *ToL*, September 26, 1972.
31. John Yudkin, *Pure White and Deadly* (London: Davis-Pointer, 1972), 84–85.
32. *ToL*, August 1, 1961; February 6, 1962; December 11, 1964; J. Yudkin and J. Roddy, "Levels of Dietary Sucrose in Patients with Occlusive Atherosclerotic Disease," *Lancet*, no. 7349 (July 4, 1964): 6–8; John Yudkin and Jill Morland, "Sugar Intake and Myocardial Infarction," *AJCN* 20 (May 1967): 503–6; Yudkin, "Sucrose and Heart Disease," *NT*, Spring 1969, 16–20.
33. *ToL*, September 22, 1961; Mark W. Bufton and Virginia Berridge, "Post-war Nutrition Science and Policy in Britain c. 1945–1994: The Case of Diet and Heart Disease," in *Food, Science, Policy and Regulation in the Twentieth Century*, ed. David F. Smith and Jim Phillips (London: Routledge, 2000), 212.
34. Yudkin was also an early advocate of the health benefits of chocolate and received funding from the Nestlé company. *ToL*, June 26, 1972; July 17, 1995; Dennis Barker and Anthony Tucker, "Squaring up to Mr. Cube," *Guardian* (London), July 20, 1995, 30.
35. John Yudkin, *Pure; Sweet and Dangerous; The New Facts about the Sugar You Eat as a Cause of Heart Disease, Diabetes, and Other Killers* (New York: Wyden, 1973).
36. *NYT*, December 21, 1974. Meanwhile, back in Britain, the previously neutral BBC distressed Yudkin with a program called "Cross Your Heart and Hope to Live," which

warned people that if they wanted to reduce their chances of being felled by a heart attack, they had to reduce their consumption of eggs, milk, butter, and cheese. *ToL*, May 27, 1974.

37. The AHA spokesperson did admit, in the fine print, that there were many other factors that contributed to heart disease, such as "heredity, sex, age, stress, personality, other diseases, cigarette smoking, inactivity, and obesity," but said that because so many of them were beyond or extremely difficult to control, they had chosen diet as "the logical candidate" as it was one of "the things that can most easily be changed." *NYT*, June 28, 1973.
38. Levenstein, *Paradox*, 193; William Dufty, *Sugar Blues* (New York: Warner, 1975), 1, 14. Dufty's book is still in print and selling well. Its latest edition claims that over 1.6 million copies have been printed.
39. Evolution does indeed seem to have endowed us with an innate "fat tooth," comparable to our "sweet tooth." Adam Drewnowski, "Why Do We Like Fat?" *JADA* 97 (1997): S58–S62; Adam Drewnowski and M. R. C. Greenwood, "Cream and Sugar: Human Preferences for High-Fat Foods," *Physiology & Behavior* 30 (April 1983): 629–33; Adam Drewnowski et al., "Sweet Tooth Reconsidered: Taste Responsiveness in Human Obesity," *Physiology & Behavior* 35 (October 1985): 617–22; Drewnowski, "Fat and Sugar: An Economic Analysis," *JN* 133 (March 2003): 838S–840S.
40. These figures are for disappearances and include the amounts used in food processing. The USDA did not break out figures for vegetable oil until 1966. USDA, Economic Research Service, "Food Availability Spreadsheets, Eggs, Margarine, Oils, Supply and Disappearances," updated March 15, 2008, http://www.ers.usda.gov/Data/FoodConsumption/FoodAvailSpreadsheets.htm#dymfg.
41. Alexander Leaf, "Observations of a Peripatetic Gerontologist," *NT*, September/October 1973, 6; Charles Percy, "You Live to Be 100 in Hunza," *Parade*, February 17, 1974, 1012; "Paradise Lost," *NT* 13 (May/June 1978): 8. Leaf published his findings in the *National Geographic* and *Scientific American* as well. His report that every day a different man was sent to run the 120 miles back and forth to Gilgit to pick up the mail set the writer and runner Erich Segal to musing that people feeling their mortality might want to run in a "Hunzathon," from New York to Boston. *NYT*, December 18, 1977. Five years later, Leaf retracted his story about Hunzakut longevity, but in the low-circulation journal *Nutrition Today*. "Paradise Lost," 8.
42. U.S. Congress, Senate, Select Committee on Nutrition and Human Needs, *Diet Related to Killer Diseases, re: Meat: Hearings*, 95th Cong., 1st sess., 1977, p. 6.
43. Sugar and salt consumption were also to be severely reduced. U.S. Congress, Senate, Select Committee on Nutrition and Human Needs, *Dietary Goals for the United States* (Washington, DC: USGPO, 1977), 12–14; Michele Zebich, "The Politics of Nutrition" (PhD diss., University of New Mexico, 1979), 172–74; Taubes, *Good Calories*, 46; Michael Pollan, *In Defense of Food* (New York: Penguin, 2008), 23.
44. Nick Mottern, "Dietary Goals," *Food Monitor*, March/April 1978, 9; Marion Nestle, *Food Politics* (Berkeley: University of California Press, 2002), 41–42.
45. Sam Keen, "Eating Our Way to Enlightenment," *Psychology Today*, October 1978, 62.
46. *NYT*, August 3, 1977.

47. USDA, ERS, Food Availability Spreadsheets, "Beef: Per capita availability, adjusted for loss/1, 1970–2007." Beef consumption also suffered from a steep decline in the price of chicken and competition from leaner pork, marketed as "the other white meat."
48. Taubes, *Good Calories*, 160–67; Michel de Lorgeril, *Cholesterol, mensonges et propagande* (Paris: Thierry Souccar, 2007), 113–14.
49. The drug was cholestyramine. Diet alone reduced cholesterol by a minute amount, from 279 mg to 277 mg. John Yudkin, "Body Politic," *ToL*, April 7, 1984; Gary Taubes, "What If It's All Been a Big Fat Lie?" *NYTM*, July 7, 2002. Two years earlier, the results were announced of a ten-year large-scale study in which 6,000 men who had their cholesterol reduced by drugs or diet, were told to quit smoking, and had their blood pressure reduced were compared with 12,000 men who received their "usual care." Much to the investigators' surprise, the difference in death rates between the two groups was insignificant. They then speculated that this was because most of the "usual care" group had heeded the advice of such groups as the AHA and Cancer Society and cut down on eating fats, smoking, and had their blood pressure reduced. *WSJ*, October 6, 1982.
50. "Hold the Eggs and Butter," *Time*, March 26, 1984. In fact, the data by no means supported this. Critics called these conclusions "unwarranted, unscientific, and wishful thinking," and "an unconscionable exaggeration of the data." Taubes, *Good Calories*, 57.
51. The estimates of the cholesterol levels that made one at risk were steadily lowered. In this case, the "at risk" population was defined as those with levels over 200 mg/dl, which is the large majority of the adult population. Southwestern Medical Center, "National Cholesterol Education Program," http://www8.utsouthwestern.edu/utsw/cda/dept27717/files/97623.html; Frederick J. Stare, Robert E. Olson, and Elizabeth M. Whelan, *Balanced Nutrition: Beyond the Cholesterol Scare* (Holbrook, MA: Bob Adams, 1989), 146; Mary Enig and Sally Fallon, "The Oiling of America," *Nexus Magazine*, December 1998–January 1999 and February–March 1999.
52. Taubes, *Good Calories*, 50; "The Soft Science of Dietary Fat," *Science* 291 (March 30, 2001): 2536. By then, experts using questionable estimates of what constituted dangerous serum cholesterol levels were warning that half of all Americans had "borderline high" or "high" levels of cholesterol in their blood, even though other experts calculated that only 16 percent of Americans could truly be considered in the "high risk" category. Stare et al., *Balanced Nutrition*, 157, 166.
53. It was expected that one-quarter of patients would be told that they had "dangerously high blood cholesterol levels." Russell L. Smith and Edward R. Pinckney, *Diet, Blood Cholesterol and Coronary Heart Disease: A Critical Review of the Literature* (Sherman Oaks, CA: Vector, 1992), vol. 2, 2–24.
54. Stare et al., *Balanced Nutrition*, 167; emphasis in the original.
55. "Searching for Life's Elixir," *Time*, December 12, 1988.
56. Patricia A. Crotty, *Good Nutrition?: Fact and Fashion in Dietary Advice* (London: Allen and Unwin, 1991), 58. In fact, the senators' wives had played a major role in converting them to lipophobia. Zebich, "Politics of Nutrition," 172.
57. The epidemiologist Meir Stampfer later said, "People looked at it and said, 'Here it is—fat causes breast cancer.' You could plot G.N.P. against cancer and get a very similar

graph, or telephone poles. Any marker of Western civilization gives you the same relationship." *NYT*, September 27, 2005.

58. The report was also particularly keen on fruits and vegetables containing vitamin C and beta-carotene as preventing cancer, something that was never supported by further research. *NYT*, June 17, June 23, 1982.
59. The diet also called for avoiding obesity and alcohol and eating more fiber. *NYT*, February 11, 1984.
60. *NYT*, January 1, 1987.
61. Stare et al., *Balanced Nutrition*. By 1990 it was also being noted that deaths from heart disease had been steadily falling, even though American fat consumption had been rising, and that since 1950 Japan, which had the greatest increase in fat consumption in the world, had also seen a steady decrease in deaths from heart disease. Dale M. Atrens, "The Questionable Wisdom of a Low-Fat Diet and Cholesterol Reduction," *Social Science and Medicine* 39 (1994): 433–47.
62. *NYT*, February 8, 2006.
63. Ancel Keys and Margaret Keys, *How to Eat Well and Stay Well the Mediterranean Way* (Garden City, NY: Doubleday, 1975); Ancel Keys et al., *Seven Countries: A Multivariate Analysis of Death and Coronary Heart Disease* (Cambridge, MA: Harvard University Press, 1980). Keys had issued preliminary reports anticipating these conclusions in 1970 and 1975. Ancel Keys, ed., *Coronary Heart Disease in Seven Countries* (New York: American Heart Association, 1970). A major problem was that collaborators in different places did not use the same ways of calculating diet and health outcomes, making conclusions drawn from comparisons very suspect. Russell Smith, a statistician, wrote, "The dietary assessment methodology was highly inconsistent across cohorts and thoroughly suspect. In addition, careful examination of the death rates and associations between diet and death rates reveal a massive set of inconsistencies and contradictions. . . . It is almost inconceivable that the Seven Countries study was performed with such scientific abandon." Smith and Pinckney, *Diet, Blood Cholesterol and Coronary Heart Disease*, 4–49.
64. In a further reordering of the fats table, olive oil's mono-unsaturated fats were said to be much better for the heart than the polyunsaturated ones in other vegetable oils.
65. Michael Symons, "Olive Oil and Air-Conditioned Culture," *Westerly* 4 (Summer 1994): 27–31; Patricia Crotty, "The Mediterranean Diet as a Food Guide: A Problem of Culture and History," *NT* 33 (November/December 1998): 227–32; Dun Gifford, "The Mediterranean Diet as Food Guide: A Comment," ibid., 233–43; Crotty, "Response to K. Dun Gifford," ibid., 244–45; Christine Wilson, "Mediterranean Diets: Once and Future?" ibid., 246–49. In the interests of full disclosure, I should admit to having been paid to speak at two of these conferences, in Boston and Toronto. However my message, which was to pooh-pooh food and health claims and to suggest that this too, like all else, would pass, was not particularly well-received by the organizers, and I was not invited to any of the other conferences, in more exotic locales such as Rome, Madrid, and Tunis.
66. The results were published in English early the next year as Serge Renaud and Michel de Lorgeril, "Wine, Alcohol, Platelets and the French Paradox for Coronary Heart Disease," *Lancet* 339 (June 20, 1992): 1523–26. *60 Minutes* ran the program again in 1992 and 1993 and did a follow-up report in 1995.

67. Renaud and de Lorgeril, "Wine, Alcohol, Platelets," 1523–26; De Lorgeril, *Cholesterol*, 114.
68. Michel de Lorgeril et al., "Mediterranean Diet, Traditional Risk Factors, and the Rate of Cardiovascular Complications after Myocardial Infarction," *Circulation* 99 (February 16, 1999): 779–85; De Lorgeril, *Cholesterol*, 116–19.
69. *NYT*, March 23, 1999. The lipophobes did face a more serious challenge closer to home, where the popular Dr. Robert Atkins was denying the dangers of saturated fat. However, they were handed a gift in 2003 when, at age seventy-two, Atkins slipped on the ice outside his New York City home, hit his head on the sidewalk, and died. They quickly spread the rumor that he had died of a heart attack, something they supported with purloined medical records showing that he had serious heart problems. *NYT*, February 11, 2004.
70. Ancel Keys, "Mediterranean Diet and Public Health: Personal Reflections," *AJCN* 61 (Supplement, June 1995): 1321S–1323S; Nancy Harmon Jenkins, *The Mediterranean Diet Cookbook: A Delicious Alternative for Lifelong Health* (New York: Bantam, 1994).
71. *Daily Telegraph* (London), November 8, 2008; "Le poids des Européens: Les Françaises les plus minces et les plus insatisfaites!" Observatoire CNIEL des Habitudes Alimentaires (OCHA), *Actualités*, April 2009, http://www.lemangeur-ocha.com/; Claude Fischler and Estelle Masson, *Manger: Français, Européens et Américains face à l'alimentation* (Paris: Odile Jacob, 2007), 24–25.
72. CreteTravel.com, http://www.cretetravel.com/Cretan_Diet/Cretan_Diet_1.htm.
73. *WSJ*, December 13, 1988; March 12, April 2, 1990; *NYT*, October 22, October 29, 1997; Gary Taubes, "The Soft Science of Dietary Fat," *Science*, March 30, 2001, 1; *Blood Weekly*, November 16, 2006, 54.
74. *NYT*, March 1, 1996; Walter C. Willett et al., "Intake of *Trans* Fatty Acids and Risk of Coronary Heart Disease among Women," *Lancet* 341 (1993): 581–85; William S. Weintraub, "Is Atherosclerotic Vascular Disease Related to High-Fat Diet?" *Journal of Clinical Epidemiology* 55 (2002): 1064–72; Sylvan Lee Weinberg, "The Diet-Heart Hypothesis: A Critique," *Journal of the American College of Cardiology* 43 (2004): 731–33; Steinberg, *Cholesterol Wars*, 197.
75. Willet et al., "Intake"; *G&M*, October 28, October 29, 2003.
76. *NYT*, July 3, 2001.
77. E. Lopez-Garcia, et al., "Consumption of Trans Fatty Acids Is Related to Plasma Biomarkers of Inflammation and Endothelial Dysfunction," *JN* 135 (March 2005): 562–66.
78. Fran B. Hu et al., "Types of Dietary Fat and the Risk of Coronary Heart Disease: A Critical Review," *Journal of the American College of Nutrition* 20 (2001): 15.
79. Steinberg, *Cholesterol Wars*, 7.
80. In 2004 the AHA joined the NIH in calling for sharply reducing the cholesterol levels for people at moderate or high risk of heart disease—something that was now widely acknowledged could only be done with statins. When it was pointed out that eight of the nine experts recommending this great increase in the number of people who should take statins had received funding from the drug companies who produced them, the response was that it would be practically impossible to find anyone who knew anything about the field who had not. *NYT*, July 20, 2004.

81. This included endorsing Vytorin's television commercials that undermined the AHA's historic diet-heart message by saying that most of the blood's cholesterol came not from diet, but from the body itself. *TV Week*, October 31, 2007; "Vytorin Ads Scrutinized by Lawmakers," NewsInferno.com, January 18, 2008, http://www.newsinferno.com/legal-news/vytorin-ads-scrutinized-by-lawmakers/; *NYT*, January 24, 2008.
82. *NYT*, January 24, March 31, April 2, September 5, 2008; "Cholesterol Lowering and Ezetimibe," editorial, *NEJM* 358 (April 3, 2008): 1507–8.
83. In 2007, with its revenues totaling $668 million, the AHA's CEO was rewarded with a salary of over $1 million. Of its revenues of $642 million in 2008, only $168 million went for research, while $131 million was directly attributable to fund-raising and $275 million was spent on "public health education," much of which was, as we have seen, was indirectly aimed at fund-raising. AHA, Financial Statements, June 30, 2003; June 30, 2007; and June 30, 2008, http://www.americanheart.org; AHA, Internal Revenue Service, Form 990, 2008, http://www.americanheart.org/downloadable/heart/12586449685482008-2009%20Form%20990%20.pdf, December 19, 2009.
84. Steinberg, *Cholesterol Wars*, 197.
85. Joel Kauffman, "Bias in Recent Papers on Diets and Drugs in Peer-Reviewed Medical Journals," *Journal of American Physicians and Surgeons* 9 (Spring 2004): 12; De Lorgeril, *Cholesterol*, 123–99; Paul M. Ridker et al., "Rosuvastatin to Prevent Vascular Events in Men and Women with Elevated C-Reactive Protein," *NEJM* 359 (November 9, 2008): 2195–207. Even the *New York Times* health columnist Jane Brody, who for years had crusaded against saturated fats, changed her tune and turned her guns on CRP. *NYT*, January 13, 2009.
86. Paul Elliot et al., "Genetic Loci Associated with C-Reactive Protein Levels and Risk of Coronary Heart Disease," *JAMA* 302 (July 1, 2009): 37–48.
87. Patty W. Siri-Tarino, Qi Sun, Frank B. Hu, and Ronald M. Krauss, "Meta-Analysis of Prospective Cohort Studies Evaluating the Association of Saturated Fat with Cardiovascular Disease," *AJCN*, doi: 10.3945/ajcn.2009.27725. The article—and an accompanying one by the same authors supporting the idea that the increased intake of carbohydrates that often accompanies reductions of fats in the diet increases the risk of heart disease—met with a vigorous rejoinder from Jeremiah Staimler, a prominent defender of the diet-heart theory. However, he did admit that "no definitive diet-heart trial has ever been done, and it is unlikely that one will ever be done." Patty W. Siri-Tarino et al., "Saturated Fat, Carbohydrate, and Cardiovascular Disease," *AJCN*, doi: 10.3945/ajcn.2008.26285; Jeremiah Staimler, "Diet-Heart: A Problematic Revisit," *AJCN*, doi: 2010:29216.
88. *G&M*, February 10, February 12, 2010.
89. *Guardian* (London), July 20, 1995; Taubes, *Good Calories*, 119–24, 324–48, 404–15. Some experiments also showed that putting a "low-fat" label on a food caused people, especially overweight ones, to underestimate its calories and to eat bigger helpings as well as to indulge themselves in other foods. Brian Wansink and Pierre Chandon, "Can 'Low-Fat' Nutrition Labels Lead to Obesity?" *Journal of Marketing Research* 43 (November 2006): 605–17.
90. "Dietary Sugars Intake and Cardiovascular Health: A Scientific Statement from the American Health Association," *Circulation* 109 (September 15, 2009): 1011–20.

91. Michael Marmot, *The Status Syndrome: How Social Status Affects Our Health and Longevity* (New York: Holt, 2004). A subsequent analysis of the data on British civil servants attributed much of the discrepancy to behavioral differences—namely, smoking, excessive alcohol intake, poor diet, and lack of exercise—rather than stress. However, an editorial commentary on it said that it showed that the time had come to move on from the stress vs. health behavior debate. It pointed out that these detrimental behaviors were often the products of stress, and that children brought up in the stress of poverty often did not develop the same capacity for "self-regulation" as those raised in higher socioeconomic strata. Silvia Stringhini et al., "Association of Socioeconomic Position with Health Behaviors and Mortality," *JAMA* 303 (March 24/31, 2010): 1159; James R. Dunn, "Health Behavior vs. the Stress of Low Socioeconomic Status and Health Outcomes," ibid., 1199–200.

Coda

1. Claude Fischler, *L'Homnivore: Le goût, la cuisine et le corps* (Paris: Odile Jacob, 1990), 368–71. A survey in Washington State in 1997–98 found little evidence of this kind of "nutritional backlash." However, one-quarter of respondents did think that eating low-fat foods took the pleasure out of eating. R. Patterson et al., "Is There a Consumer Backlash against the Diet and Health Message?" *JADA* 101 (2001): 37–41.
2. I tell this story in *Revolution*, 44–59.

Index

Note: Page references in *italics* refer to illustrations.